2ND EDITION

CRUISIN' THE FOSSIL FREEWAY

AN EPOCH TALE
OF A SCIENTIST AND AN ARTIST
ON THE ULTIMATE 5,000-MILE
PALEO ROAD TRIP

WITH PALEONTOLOGIST **KIRK JOHNSON**

AND ARTIST **RAY TROLL**

CHICAGO REVIEW PRESS

Text copyright © 2007 by Kirk Johnson

Illustrations copyright © 2007 by Ray Troll

Photograph credits:

© Kirk Johnson: pages 6 (left), 13, 14 (left), 16, 21, 33, 45, 49, 51, 61, 63, 65, 66, 69, 75, 84, 85, 91 (bottom), 96, 99, 103, 104, 110, 114, 115, 117, 124, 135, 136, 146 (all), 147 (bottom), 160, 167, 183, 184, 186, 191, 194

© Ray Troll: pages 1, 2 (all), 6 (insets), 29, 41 (all), 42, 44, 64, 83, 87, 91 (top), 107, 109, 122 (bottom), 128, 134 (all), 152, 153, 159, 164, 165, 169, 173, 175, 177, 178, 200

© DMNS, Rick Wicker: pages iv, 7, 14(right), 137, 142, 145, 147 (top), 181, 182, 187

Lyrics from "There Is a Mountain" by Donovan used by permission of Peermusic Ltd.

Published by Chicago Review Press Incorporated

814 North Franklin Street

Chicago, Illinois 60610

ISBN 978-1-64160-915-9

Library of Congress has cataloged the first edition as follows:

Johnson, Kirk R.

  Cruisin' the fossil freeway : an epoch tale of a scientist and an artist on the ultimate 5,000-mile paleo road trip / by Kirk R. Johnson and Ray Troll.

    p. cm.

  Includes index.

  ISBN-13: 978-1-55591-451-6 (pbk. : alk. paper) 1. Paleontology--West (U.S.)
2. Dinosaurs--West (U.S.) 3. Fossils--West (U.S.) 4. West (U.S.)--Description and travel. 5. United States--Discovery and exploration. I. Troll, Ray, 1954- II. Title.
  QE711.3.J64 2007
  560.978--dc22

2007019480

Printed in China

0 9 8 7 6 5 4 3 2 1

Design: Ann W. Douden

Layout for second edition: Jonathan Hahn

This book is dedicated to my parents, Dick and Katie Jo Johnson. My mom's childhood on a Wyoming sheep ranch and our annual family road trips to Casper filled my brain with stories of the West, delivered me to dozens of rock shops, and put me in front of my first free-range fossils. My dad regularly imposed his love of hiking and mountain climbing on his reluctant eight-year-old son. I remember waking one morning, high in the Olympic Mountains, and crawling out of the tent to the sight of an endless vista of receding ranges. That view, at that time, seared my mind with the expansiveness and potential of the wild world and set me on a path of exploration and excavation that continues to this day.

—K. J.

I dedicate this book to my "sole" brother Brad Matsen, just another vertebrate who first showed me the joys of creative literary collaboration and the fun and adventure to be found on the open road. I hope he will see it in his heart to forgive me for hooking up with another writer.

—R. T.

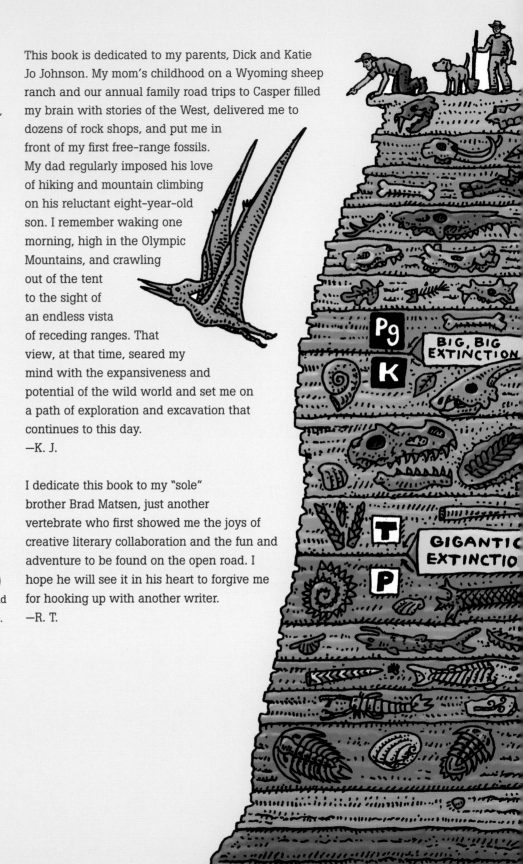

© Ray Troll 2017

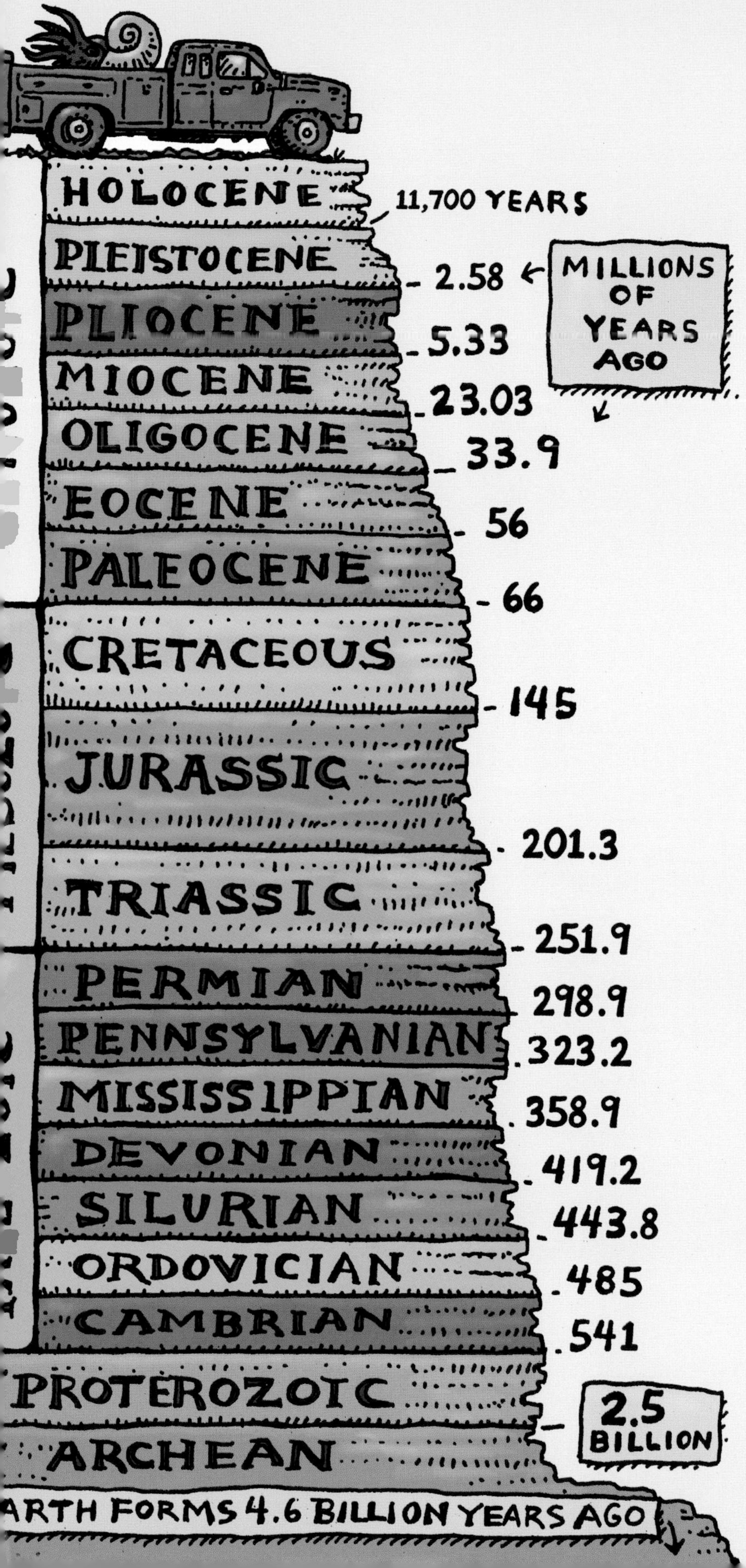

HOLOCENE .... 11,700 YEARS

PLEISTOCENE ..... 2.58 ← MILLIONS OF YEARS AGO

PLIOCENE 5.33

MIOCENE 23.03

OLIGOCENE 33.9

EOCENE .. 56

PALEOCENE .. 66

CRETACEOUS .. 145

JURASSIC .. 201.3

TRIASSIC .. 251.9

PERMIAN 298.9

PENNSYLVANIAN 323.2

MISSISSIPPIAN 358.9

DEVONIAN 419.2

SILURIAN 443.8

ORDOVICIAN 485

CAMBRIAN 541

PROTEROZOIC 2.5 BILLION

ARCHEAN

EARTH FORMS 4.6 BILLION YEARS AGO

TABLE OF CONTENTS

PREFACE TO THE SECOND EDITION

Ray and I published the first edition of this book in 2007. It was based on a series of road trips that we took between 1997 and 2005. We used film cameras, faxes, pay phones, CDs and a boom box, and drove a truck that guzzled leaded gasoline. By the time the book came out, we had been at it for nine years. We immediately fell into paleopostpartum depression and started talking about a second book. That led to another decade of road trips between 2009 and 2017. This time, we used iPhones, digital cameras, rental cars, float planes, and a fleet of different boats. This resulted in *Cruisin' the Fossil Coastline* in 2018. Both books, birthed in wanderlust, transformed themselves into traveling exhibits. By 2019, we were once again struck by the hunger to get back on the road and meet some new fossils. We talked about doing a book on the fossils of the East Coast and even made plans in March 2020 to participate in the excavation of a fossil rhinoceros in Florida. Then the pandemic hit, and the world paused.

We have been chasing these stories for the last 26 years, and over this time, many of the fine people we visited have passed away. Meanwhile, new fossil lovers have been born, new paleontologists have been minted, and the continent continues to spew out amazing and surprising new fossil discoveries. The best stuff truly is still in the ground.

In 2010–2011, I led a surreal 69-day excavation in Snowmass Village, Colorado, that yielded more than 5,426 Ice Age bones—a literal herd of mammoths, mastodons, and other animals. In the following decade, the Denver Museum of Nature and Science grew into a paleontology powerhouse. Its teams uncovered dozens of dinosaurs in the San Juan Basin of New Mexico, the Grand Staircase Escalante badlands of Utah, the Denver Basin of Colorado, the Williston Basin of North Dakota, and the Powder River Basin of Wyoming. In 2015, Denver curator Tyler Lyson found a trove of fossil mammals from the first million years following the extinction of the dinosaurs just minutes from downtown Colorado Springs. And in 2017, construction workers in urban Colorado even found a nearly complete *Torosaurus* dinosaur skeleton while building a firehouse in downtown Thornton.

In 2016, a cowboy in Montana found near perfect skeletons of a *Triceratops* and a young *T. rex* that appeared to be locked in permanent mortal combat. The bar owner that bought our Evolvo is now the junior senator from Colorado. And the K-T boundary is now known as the K-Pg boundary. And speaking of the K-Pg boundary, in 2015 a couple of commercial paleontologists found a logjam of fossil fish that turned out to be the remains of a vat of quicksand that formed within minutes of the dinosaur-killing asteroid impact.

In 2012, I left my job as the chief curator of the Denver Museum of Nature and Science to become the Sant Director of the Smithsonian's National Museum of Natural History. In 2019, the museum opened the David H. Koch Hall of Fossils-Deep Time, the story of life on Earth on the national mall. And in 2022, Ray bought a building in Lindsborg, Kansas, and entered the museum-building phase of his life. Apparently, fossils never sleep, so we thought it might be a ripe time to buff up a new and improved second edition to *Cruisin' the Fossil Freeway*.

SABER-TOOTHED EVERYTHING

I woke with a jolt, soaked in sweat, in a stinky little cabin on a riverboat on the upper Amazon in Peru. Snarling animals with slashing saber teeth snapped at my heels, the remnants of a horrible dream. Stocked with images of all of prehistory, my paleontologist's brain concocted appalling saber-toothed cats and tigers, saber-toothed marsupials, saber-toothed uintatheres, saber-toothed deer, and a staggering melee of other extinct creatures with saber teeth but no common names. I was 36 years old and suffering from another Lariam nightmare.

Lariam is a drug used to prevent malaria, but it is also renowned for its side effect of appallingly vivid slasher nightmares. Many users report murderous dreams where family members set on each other with knives or friends chop each other apart with machetes. Suffering a similar fate, I heard the shrill "wree, wree, wree" from the shower scene in Hitchcock's *Psycho* that morning, only my Janet Leigh had much bigger teeth.

As I shook off my terror and pondered this nightmare, two realizations suddenly hit me: walrus are saber-toothed seals, and only Ray Troll could paint my dream.

***

I first met Ray at the Burke Museum in Seattle in 1993, but the meeting had been destined for a long time. Ray is a fish-obsessed artist whose work is densely packed and jarringly unique. Early in his career, he found that it was easier to sell art if it was on a T-shirt, and his images can now be seen on more than a million chests up and down the West Coast. Growing up in Seattle, I had spent my early 20s wearing those T-shirts. Ray remembers looking up from a lobe-fin fish–filled museum case that day to see a hulking 6′ 3″ frame bearing down on him. I was

really excited to meet the man whose "Spawn 'til you Die" and "Humpies from Hell" shirts filled my closet. Ray felt like he was about to get mugged.

Despite our shared interests, it wasn't until my 1996 Lariam dream that I realized I simply had to work with Ray. But how could I lure this artist into working with me? I knew that for the last 20 years he had lived in Ketchikan, a small fishing, logging, and tourist town on Revillagigedo Island at the southern tip of Alaska, and I figured it would be best if I went to him. So, I tacked a few extra days onto a fishing trip and found myself in a phone booth in rainy Ketchikan, cold-calling the Troll.

I didn't really have a plan for my pitch, but I was aided by the fact that Ray has a curious habit shared by artists who dabble in the realm of natural history: he collects scientists. Between that and the fact that people in Ketchikan don't get too many drop-in visitors, it wasn't much of a sales job after all. Ray answered the phone and invited me to the Soho Coho, his smart little art gallery on a boardwalk over a salmon stream. Our conversation went well, and we ended up in his studio talking about fossils and fish. I told

The Soho Coho gallery on a boardwalk over a salmon stream.

1

him about the Lariam dream and the fishy wonders of the Amazon, which I visit each year for an immersion in tropical biodiversity. By the end of the conversation, Ray had signed up for my next Amazon trip and we started plotting projects together.

But let me back up and start at the beginning of this whole paleo obsession, at least my beginning. When I was five, there was a pretty redhead next door who owned a rock polisher, one of those metal cans on rubber rollers hooked up to a small engine. You fill the can with rocks, grit, and water, turn it on, and let it slowly rotate for a week. Then you change out the coarser grit for finer and repeat the process. Three times you change the grit, with the whole process taking a month. A month is an insanely inscrutable amount of time for a five-year-old. Kelsey, a mature nine-year-old, had the patience of a saint, and I remember clearly that sunny morning when she called me over to her porch. The month was finally up, and she let me watch as she lifted the can off the tumbler. She opened it and poured out a muddy slurry. I was stunned as she washed away the slimy polishing compound to reveal a trove of gorgeous glistening agates. It was one of those perfect Seattle summer mornings, and I remember the sun backlighting Kelsey's red hair and bouncing off the polished red agates. I fell in love with rocks, and girls, at that moment.

It took me many more years to even begin to get a grip on the girl thing, but the rocks got traction immediately. I pored over the gravel in driveways and on beaches. I planned and pouted so my parents would interrupt family road trips to stop at rock shops. My dad caved quickly to my incessant pleading for a rock tumbler, and before long I had my own tumbling drums of grit and gravel. I discovered the 500 section of the local public library and was soon hauling home stacks of books about rocks, gems, and fossils.

Memorable stones began to present themselves to me. A fossil leaf on a trail at summer camp near Mount Rainier. A brachiopod from the top of Casper Mountain that I thought was a fossilized rattlesnake tail. A chunk of black limestone from the Canadian Rockies that was patterned

with stark white bryozoans. Then, a little later, beautifully pyritized ammonites from the beach gravels of Lyme Regis on the Dorset Coast of England. By the time I was 12, I was a goner, hooked for life on this strange pastime of seeking and hoarding stones. Fossils owned me, and I owned a lot of them. It was around this time that someone showed me one of the fossil crabs from the Olympic Peninsula. These are still some of the most amazing and precious fossils I have ever seen. Crabs, complete with all their legs and claws, preserved, sometimes with the original shell color intact, in tight round concretions. Little round tombs with perfect crustaceans. I had to have one. I had to find one. Unlike most kids on the planet, I didn't give much thought to dinosaurs. Instead, I was passionately hooked on fossils that I thought I might have a chance of finding.

In time, probably not much time, friends of friends steered me to a man who knew how and where to find fossils: highway department maintenance man Bill Buchanan. His house was only a five-hour drive from ours. My dad and I drove to Clallam Bay on the north coast of the Olympic Peninsula to his smoky cottage on the edge of town. He was a generous man who gave me fossils as well as information about how to find them.

Under his guidance, I learned how to walk down a rain-soaked, rocky beach with an eight-pound sledgehammer, smacking likely boulders for the treasures they held. Bill cracked open a soccer ball–sized concretion that contained a perfect crab on our very first venture together. He didn't even hesitate to give me the precious rock. That's a kindness I remember with great clarity. In time, I too had a personal armory of sledgehammers and a keen eye for just the right kind of rock.

It was about this time that I met Wes Wehr, an impossibly quiet artist with an abiding passion for petrified wood. He had already worked his way into the Burke Museum as an unpaid curator of fossil plants and was busy collecting correspondence from distant paleobotanists. Bill Buchanan had found some fossil conifer cones in the crab nodules and, through me, Wes connected Bill with Chuck Miller, a professor at the University of Montana, who specialized in fossil cones. Wes and I began to gather cones, and eventually Chuck named a fossil cone after Bill. Wes was a city artist who didn't know how to drive a car. The summer after I got my driver's license, he and I headed off in my parents' orange Audi for a weeklong drive across eastern Washington and Oregon to locate fossil sites that we had read about in library books. Early in the trip, we struck pay dirt in the little gold-mining town of Republic in northeastern Washington. A forgotten Eocene lake bed was exposed in the hills around town, and we had blundered around asking the locals if they knew where to find fossils. They didn't, and we weren't having any luck ourselves, so we decided to leave. Our car was parked on the main street at the south end of town. As I walked around the back to get in, I kicked a little piece of roadside shale. To my amazement, the rock

fell open, revealing a stunning little sprig of dawn redwood foliage. We quickly aborted our departure and dug into the drainage ditch. The buff layers immediately began to yield beautiful fossil leaves, cones, insects, and flowers. On subsequent trips to this spot, we found so many perfect fossil flowers that I started giving them to girls I liked, and their sweet response made me realize that plants are the best kind of fossils.

We stopped at the Oregon Museum of Science and Industry in Portland on the way back to Seattle, where we met a 21-year-old paleobotanical wunderkind named Steve Manchester. He was in a back room supervising a team of six high-school kids who were using Elmer's glue to reassemble shattered fossil tropical rain forest leaves from eastern Oregon. Some of the leaves were in 20 pieces, and I was amazed at how diligently these kids struggled over their rocky jigsaw puzzles. I was more surprised, and a little bit dismayed, to realize that I wasn't the only kid in the world who thought about fossil leaves. I didn't realize it at the time, but this road trip sealed my fate and destined me to become a paleobotanist.

A few years later, as an interested but aimless college junior studying art, geology, and rugby at Amherst College in central Massachusetts, I accepted a summer research assistantship that landed me in the tiny town of Marmarth in the southwestern corner of North Dakota. My stated job was to measure coal seams and trace them along hillsides to see how they thickened and thinned. The 65 million-year-old coal is soft and brown, and it lies buried in layers of sand and clay that just never got buried deeply enough to turn into rock. I spent the summer, much of it alone, wandering through a maze of buttes and gullies with a shovel, digging trenches through the shallow prairie soil to expose the coal seams. Sometimes the holes yielded fragments of fossil leaves. The utter remoteness of place and my daily grind of digging into the earth gave me an appreciation for the vastness of space and time. It also made me realize that it was pointless to like fossils without understanding geology.

Those months on the High Plains reawakened childhood memories of driving from Seattle to Wyoming to visit the ranch where my mother spent her childhood. I was rediscovering the things that I loved about the plains: the huge sky where a cast of clouds plays out the continuous, overwrought drama of real weather; the smell of ozone that mixes with intense sage when a thunderstorm is imminent; the eternal search for arrowheads and rattlesnakes. That summer, I saw a tornado, I found arrowheads, and I grabbed a rattlesnake, but mostly, I realized that I could travel through time with a shovel. I began to understand Faulkner's lines "The past is not dead. In fact, it is not even past." By the end of August, I had converted to geology and to North Dakota with the fervor of the born again.

Ray took another path. He grew up in the dinosaur frenzy of the late 1950s and, like every other kid on the planet, he was dinosaur obsessed. But unlike the other kids, he was really good at drawing the dinosaurs. As a nascent artist, his very first crayon drawings were of snarling *T. rex* and cowering, bloodied *Triceratops*. The first word he wanted and managed to spell was *dinosaur*; by the age of six, he knew the Latin names for dozens of them. Making sound effects that worried his mother, he spent countless hours drawing prehistoric animals. He bought every dinosaur toy available in the 1960s and collected cereal-box treasures and postcards of prehistoric creatures. He was one of those kids who found oddly shaped rocks in the playground and adamantly declared that he had discovered rare dinosaur bones. As a museum curator, I

see those kids every week, and it is my job to let them down easily as I explain that their dinosaur egg is a rounded chunk of granite. But aside from a lucky brachiopod found in his grandparents' driveway, Ray didn't collect his first real fossil until he was 40 years old.

Instead, he grew up as a fossil-obsessed but fossil-deprived military brat, bouncing from air force base to air force base. He had an innate talent for seeing and drawing, and his early paintings were elaborate battle scenes with casts of thousands. By the time Ray was in high school, his family had settled in Kansas, and he soon became aware of the phenomenal fossils of the Sternberg Museum in Hays, on the other side of the state. Fruits of the labors of a whole fossil-finding family, the collection at the Sternberg is beyond compare. Here, in an unremarkable town on the plains of western Kansas, are the remains of the fish, reptiles, and birds that swam in and flew over a huge salty sea that covered Kansas 85 million years ago. Whole sharks, mosasaurs, plesiosaurs, pteranodons, and giant fishes lived where today you only see farm animals. When he was in high school, Ray and his pals road-tripped to the chalk beds of western Kansas to find their own fossils, but he didn't know how to look for them, so he didn't find any. Not long after, Ray hooked up with a retired schoolteacher who was making educational filmstrips, and soon he was being paid for his drawings of prehistoric creatures.

Ray was born in 1954, so technically he was part of the '60s. But the truth is that the '60s didn't get to rural Kansas until the early '70s, so Ray came of age just about the time the funk rolled into town. He went to a local college and studied art, photography, and rock and roll. Avant-garde images and electric guitar riffs were added to the battle scenes and dinosaurs in his head and, surprisingly, it all stayed in there. Graduate school and a stint in Seattle buffed off his midwestern edges and morphed him into an urban hipster who played in rock and roll bands and camouflaged his love of fossils.

Ray's move to Ketchikan in the early '80s dumped him in the epicenter of the salmon fishery and into the middle of the Haida and Tlingit Indian art revival. By the time I met him, Ray was 39, married with two kids, balding, and cranking out some amazing art. Best of all, he had recently reconnected with his childhood love of fossils.

A HARSH LESSON: No matter how hard young Raymond imagined it to be, the rock was simply NOT a fossil bone.

DANCING TO THE FOSSIL RECORD

Art and Fossils

Throughout the brief few hundred years of the discipline of paleontology, paleontologists have often worked with artists or were artists themselves. In a world of stony fragments, old bones, and flattened leaves, there's a real need for people with artistic talent, imagination, and the ability to bring lost worlds back as images. Museums and books are full of paintings of dinosaurs and their worlds.

As a curator, I have advised many artists as to what they could, and could not, legitimately include in prehistoric scenes. In these collaborations, our goal is always to create accurate, plausible, and realistic landscapes from deep time. For me, a wannabe artist, it's great fun to direct the conception of an image, to tweak its contents, and yet still be surprised by how the final image so often feels like a real place.

Ray is a different kind of paleoartist than I was used to. He's no photo-realist, he's a scientific surrealist. His art, while often paleontological, is infused with the rest of his life. In his images, extinct animals visit the modern world in daydreams, as if underground cartoonist R. Crumb time-traveled to the Cretaceous.

Monument Rocks (also known as the Kansas Pyramids) of Gove County, Kansas, are made of layers of Cretaceous chalk. These stacks of fossil plankton are full of marine fossils that lived when Kansas was at the bottom of a sea.

The Troll family explores the chalk.

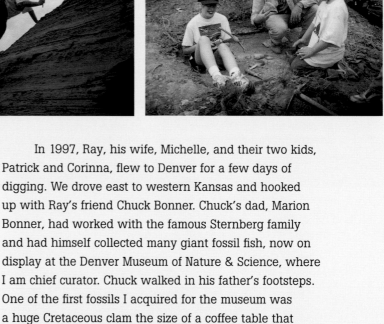

In 1997, Ray, his wife, Michelle, and their two kids, Patrick and Corinna, flew to Denver for a few days of digging. We drove east to western Kansas and hooked up with Ray's friend Chuck Bonner. Chuck's dad, Marion Bonner, had worked with the famous Sternberg family and had himself collected many giant fossil fish, now on display at the Denver Museum of Nature & Science, where I am chief curator. Chuck walked in his father's footsteps. One of the first fossils I acquired for the museum was a huge Cretaceous clam the size of a coffee table that came from Chuck. I had driven out to pick up the massive mollusk and asked Chuck if he would show me where he had collected it. "No problem," he replied, "it's just over the hill." Ten minutes later, I was standing at the edge of a little valley where the prairie had eroded away to expose a half-acre of flat-lying Cretaceous chalk. From where I stood, I could count a dozen of the meter-wide clams just lying there, waiting to be collected. I began to understand the charms of fossiling in Kansas.

The Troll family and I had a great time visiting the Bonners, who live in a small house far out on the prairie next to an old stone church full of Chuck's fossil finds. Our expedition to the Cretaceous chalk beds was a blast, but

Ray seemed unable to unearth a single fossil. His daughter found a great pectoral girdle of a big mosasaur, and the rest of us were finding pieces and chunks of ancient fish and handfuls of sharks' teeth, but Ray kept getting skunked. Finally, he sat down in exasperation near the truck and gave up. I wandered over to harass him. The key to finding a fossil is knowing what a fossil looks like. It's about shape, color, and texture, and recognizing the anomalous fragment of biological form. Often, simply remaining motionless and concentrating on a single piece of ground is all it takes. With that piece of advice, I bent over and picked up a beautifully sharp and shiny *Squalicorox* shark tooth that was about six inches from Ray's butt. It was clear that I had my work cut out for me with this fossil-obsessed but fossil-challenged artist.

By 1999, Ray and I had installed a museum exhibit called *Cruisin' the Fossil Freeway* at the Denver Museum. It's a traveling show of Ray's art with a lot of cool local fossils from the museum's collections. As part of this exhibit, we bought an old Volvo station wagon and completely worked it over with fossils, evolutionary paraphernalia, and put Charles Darwin in the driver's seat. Ray called the tricked-out car an Evolvo, and that styling little fossil car started us thinking about taking a real fossil road trip around the American West. Two years later, we were making our dream a reality.

Ray Troll and a team of volunteers painting a giant ammonite mural in Denver in 1999.

CRUISIN' THE FOSSIL FREE-WAY

I DIG DINOS

Our idea for a big fossil road map preceded

our idea for a big road trip, but they quickly grew together. Ray jump-started the map by gluing a gigantic sheet of paper to the wall of his studio in Ketchikan and tracing some state lines on it. The shape of the paper was arbitrary, and it dictated which states made it onto the final map. Our original idea was to do Colorado and Wyoming and all the states that touched them, but, because of the shape of the paper, we ended up shorting Arizona and New Mexico and adding a lot of real estate in Washington, Oregon, Nevada, and California. Each image on the map is based on an actual fossil from that spot. Ray, ever the fan of roadside eats, hid a drawing of a cheeseburger in every state.

The drawing took a solid nine months of Ray's time, with me working feverishly to supply accurate data to fill all the spaces. Hundreds of phone calls, faxes, and e-mails flew between our respective work spaces in Denver and Ketchikan. When the map was half done, I flew to Ketchikan to visit Ray. My first viewing lasted more than an hour. Despite the incredible density of prehistoric images, I was struck by the fact that fossils are so abundant, diverse, and widespread that we could have drawn dozens of different versions of this map, each populated by a completely different cast of characters. When the drawing was finally completed, Ray carefully cut the original in two to have it scanned at a local print shop. Terry Pyles, one of Ray's Ketchikan art pals, added digital color over the next few months. After a year's time, we had our map.

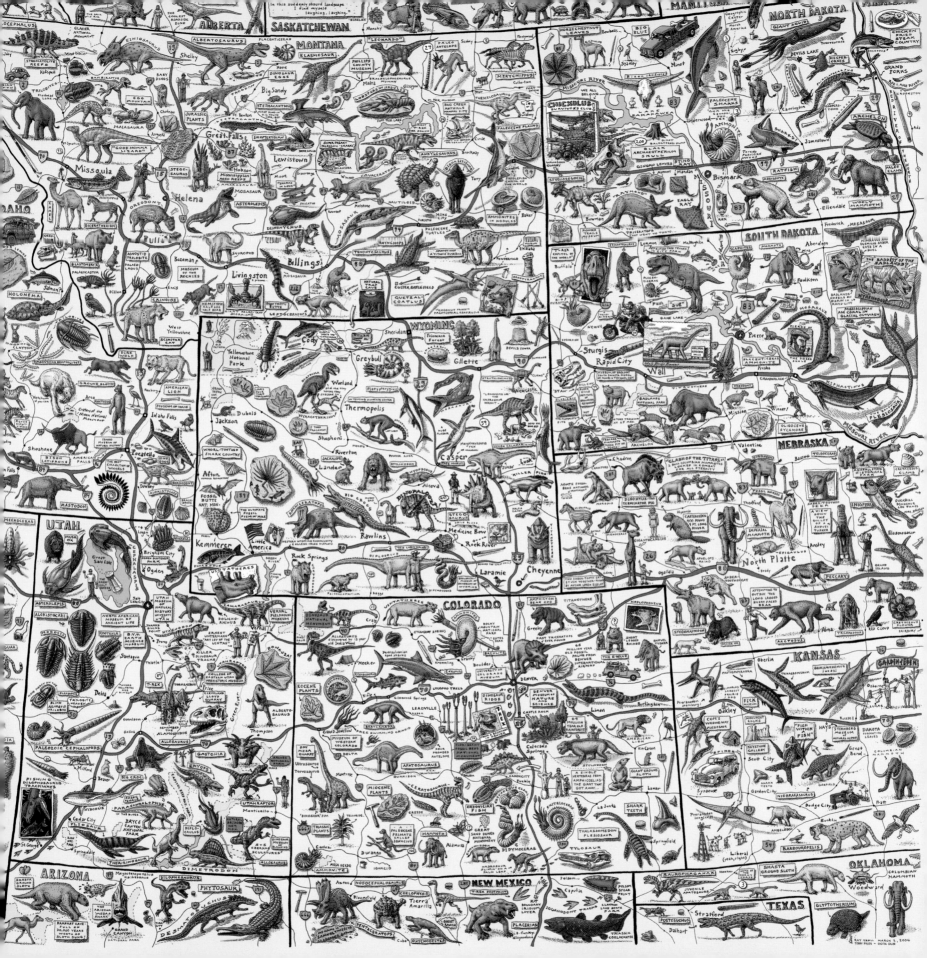

1
SUBURBAN REX

I found Ray at the baggage claim at Denver International Airport. He was sporting a jean jacket with a Haida Indian design, black jeans, shiny black shoes, and carrying a briefcase decorated with trilobites and dinosaurs. Not my idea of field clothes, but urban artiness is apparently tough to shed. Eventually Ray's huge duffel spewed onto the carousel. I remembered that I had asked him to bring the Troll family camping tent to keep our costs down, since we were going to be traveling on the cheap. We each grabbed an end of the bag, which felt like it held a corpse, and hauled it out to the parking garage to the museum's pickup.

A dark blue Ford F-250 with a partial crew cab, aluminum toolboxes, and a Tommy lift, Big Blue was purchased new into the Denver Museum's vehicle fleet back in 1983. Big Blue has an engine too small for its big body, a very finicky clutch and stiff gearshift, and a big steel tailgate that makes it nearly impossible to see out the rearview mirror. For these reasons, most of the other museum employees hated driving it, and those who did could often be seen struggling with the bucking clutch or backing the truck into unseen obstacles. Mainly by default, it became my truck, and between 1991 and the moment I picked up Ray, Big Blue and I had visited more than 500 fossil sites in a dozen western states. Like the big kid I am, I delighted in the comic possibilities of the hydraulic tailgate, often lowering it to just the right level so that I could tie my bootlaces or roll a shopping cart full of groceries onto it to load into a cooler in the bed of the truck. Big Blue is a fossil-killing monster, and it seems only right that this beast would be our time-traveling machine.

As we rolled out of the parking garage and onto the flat plains of eastern Colorado, I began to explain the wonders of the land we were driving across on our way toward town. Today, Coloradoans are pretty comfortable with the concept that they live in a place that's a mix of mountains and plains, forests and prairies, cities, small towns, strip malls, and ranch land. But it doesn't take too much memory or too many generations to get back to the time before Lewis and Clark. We imagine endless herds of bison carpeting the plains and wish we could see how it used to be. But this is a slippery slope, because "used to be" isn't a destination, it's a journey. The world of paleontology allows us to slide that slippery slope all the way back to the beginning of life on Earth. Fossils are the remains of those days, and the fossils allow us to rebuild and understand those ancient and extinct ecosystems.

A plaster-covered mammoth tusk at a construction site 10 miles south of Denver.

The Rocky Mountain West is the world's finest fossil field, home to *Tyrannosaurus rex*, *Apatosaurus*, *Allosaurus*, *Stegosaurus*, and countless other well-known dinosaurs. But that's just the beginning. The rocks and road cuts of the Rocky Mountain states contain fossils that hark back more than 500 million years. The West has not only dinosaurs but trilobites, giant beavers, American cheetahs, spiny clams, spiky plankton, killer pigs, shovel-tusked elephants, palm fronds, mammoth camels, rain forest primates, six-foot-wide ammonites, tens of billions of fossil leaves, and myriad plants and animals that haven't even been named yet. The region is bursting at the seams with cool fossils.

Some great finds have been made in and around the sprawling strip-mall city of Denver, where fossils have been known to show up in suburban backyards, on city street corners, and even at the airport. Denver was founded on the banks of the Platte River on November 22, 1858, a year and two days before Charles Darwin published *The Origin of Species*, but four years after Ferdinand Hayden found America's first dinosaur fossil on the banks of the Missouri River. On March 26, 1877, a man named Arthur Lakes was searching for fossil leaves near Golden, Colorado, when he came across a fossil bone the size of a log. He had found the first sauropod, or long-necked dinosaur, an animal that would come to be known as *Apatosaurus ajax*. Sauropods, the largest land animals this planet has ever seen, are awesome, enigmatic reptiles, deep-time denizens utterly different from anything alive today. Lakes's beast still tips the scales as one of the largest sauropods ever found. His discovery was followed that same year by even better dinosaur finds in Cañon City, Colorado, and Como Bluff, Wyoming. The western bone rush was on.

Ten years later, William Cannon, wandering up a gully about a mile west of downtown Denver, found a pair of long fossil horns. He shipped them to Othniel C. Marsh at Yale, who named them *Bison alticornis*, a new species of fossil bison. It only took two years and a more complete skull from Wyoming before Marsh realized that Cannon's bison horns were actually the business end of a *Triceratops*, and downtown Denver was credited with the world's first-known horned dinosaur.

It's been like that in Denver ever since. Someone walking around town or an alert backhoe driver digging a ditch finds a bone, and sooner or later somebody else realizes that it's a chunk of some huge extinct beast. The vaults at the Denver Museum of Nature & Science are full of these urban treasures. There's a fossil camel from the corner of 6th and Clermont, a group of Ice Age peccaries from the excavation for the Colorado capitol building, chunks of mammoths from the stream valleys, a dinosaur rib from home plate at Coors Field, and fossil palm trunks from all over town.

As I drove west, I reminisced out loud about my connection to a typical Denver fossil find. When I moved to Denver in 1990 to begin working at the Denver Museum, the city was building the new Denver International Airport (DIA), its completed white outline now fading in Big Blue's rearview mirror. One day in early spring, workers called the museum to report that they were finding giant fossil fish, or at least the tails of giant fish. When I rushed out to the site, I found that the fish tails were actually giant fossil palm fronds: the new airport was being built on a very old swamp! I watched helplessly as one grader buried a single frond that was 11 feet long and 7 feet wide. It was my second month on

Les Robinette, a retired wildlife biologist whose Denver basement is full of fantastic fossil jaws and skulls.

Museum volunteer John Shinton with a fossil palm from Denver International Airport.

the job, and I didn't have the equipment or the chutzpah to collect the 20-ton fossil.

Two years later, the museum again received a call: a guy named Charles Fickle was walking his dog through a half-built neighborhood in the suburb of Littleton when the two of them crossed a lot that had recently been scraped clean to make way for a new home. Halfway across the lot, they found a huge bone sticking out of the ground. Fickle maintains that he saw the bone first, but I'm convinced that his dog beat him to it. Fickle ran home and got his truck, which he parked over the bone to protect it until museum staff arrived. Fickle's bone was associated with many more bones, and when the dust had settled, the museum team had excavated the entire right leg, 10 teeth, a shoulder blade, and a tail vertebra of a *Tyrannosaurus rex*. At the time, this was the fifteenth *T. rex* skeleton known in the world, but it was the first with its own street address.

What becomes apparent when you start to look at all these fossils is that the Rocky Mountain region has a long history, and a lot of that history is just lying around waiting to be bumped into. With that in mind, and our stomachs grumbling, Ray and I decided to stop for a bite of breakfast at the Walnut Café and contemplate the many possibilities of our trip.

Ray was up to speed on geologic time, evolution, and extinct critters, but he had a lot to learn about other fossils, such as my beloved and much ignored plants. And when it came to rocks and geology, he was a babe in the woods. The waitress dropped off the menus as I explained to Ray, "If you want to find fossils, you've got to go where they are, you've got to understand geology. If you can understand the geometry of geology, then you're well on your way to being a successful fossil finder."

Without even cracking the menu, I decided to use pancakes to explain the idea of a geological formation. Food makes excellent geology. I once used a plate of mashed potatoes and gravy to explain plate tectonics to a table full of millionaires on a ship in Antarctica. I cut a Gondwana-shaped blob of potato into smaller pieces and pushed them away from each other with spoons.

Oceans of gravy filled in between my potatoey southern continents, and soon I had created a credible polar view of Antarctica, Australia, Africa, India, and South America. Then I ate the caloric continents. The six-pack of CEOs seemed amused, and I realized that teaching geology is easy because everybody needs to eat.

This is especially true for Ray. He loves his three squares. I like my meals too, but I'll often overindulge in the first round and sit out the second. But with Ray, before we even packed the truck for our trip, I knew I would be eating a regular breakfast, lunch, and dinner every day, and always in the most authentic local eatery we could find.

By the time my stack of walnut pancakes and cup of coffee had arrived, the lesson had begun. As I told my fossil-finding partner, the key to understanding fossil-bearing rocks is to think about familiar layers. Think about stacks of pancakes, trays of lasagna, Dagwood sandwiches, and platters of baklava. Think about your undershirt, chamois shirt, sweater, and raincoat. Think about painting your bedroom with three coats of paint. Think about an Oriental rug store where the finely woven carpets are stacked in

Giant tilted slabs of Pennsylvanian Fountain Formation at Red Rocks Amphitheatre, the best place in the world to see a concert.

tall piles, and the one you want is always at the bottom. In all of these cases, the first coat, the first layer, the first carpet, the first pancake is the lowest one. Fossil-bearing rock comes in layers, big layers, which geologists group into formations. I like to think of a geological formation as a big stone pancake.

Formations, like pancakes, have a certain thickness and a certain lateral dimension. The walnut pancakes were about half an inch thick and nine inches in diameter. I was going to have a tough time eating this demonstration. But there is no typical forma-tion; they range from 10 to 10,000 feet thick and are from 3 to 3,000 miles in diameter. They are rarely as round as a pancake, and they sometimes grade into each other at their edges. Like pancakes, they come in stacks, with the oldest ones at the bottom of the stack. I sliced my stack of pancakes in half with a knife and showed Ray the stack of formations. A similar thing happened when canyons were cut into the Rocky Mountains, exposing stacks of layers of rocks.

But unlike pancakes, which are uniform inside, the formations themselves are made of thinner layers. So maybe a piece of baklava is a better analogy for a forma-tion than a pancake. Like baklava, formations are named for the place where they were first described. Take the Chugwater Formation, first described near Chugwater Creek, Wyoming. The Chugwater Formation is about 1,000 feet thick and stretches over most of Wyoming. It's bright red and easy to distinguish from the layers above and below it. It's composed of sandstone and mudstone—rocks, by their name, that you can tell used to be sand and mud.

This sediment was deposited by streams and rivers about 230 million years ago, when all continents were connected, forming the mother of all continents: a landmass known to geologists as Pangaea. Since it takes a flat place to make flat layers (remember, it's a flat griddle that makes a pancake flat), we surmise that the Chug-water was deposited on a flat surface. The very fact that most of the forma-tions of the Rocky Mountains are composed of flat layers is enough to tell you that this region was primarily a flat place for much of its history. Of course, it's not flat now, so it's hard to imagine it that way.

The stack of rocky pancakes is called a geological column. That's the whole pile of layered rocks. In some places, because of uplifting mountains and the resultant tilt of the layers, it's possible to see the whole column at the surface. At other places, the surface of the Earth is the top of the stack and the rest of the pile is deeply buried. The Rocky Mountain region contains many examples of both. You can use the surface exposures to understand and predict the buried ones. The oil and gas industry has been doing just that for the last 100 years.

Now we were getting to the good part. Because different fossils are generally found in different formations, the rocks are the most important clue to understanding how to find fossils. Sitting there enjoying our third cup of coffee, Ray and I found ourselves in the middle of the Denver Basin, directly on top of nearly 12,000 feet of layered rock that geologists had grouped into 11 forma-tions. A hole drilled straight down from the Walnut Café would penetrate all these formations. The layer just below

ROCKS DON'T LIE

the building is about 50 feet thick and 12,000 years old. It's called the Cherry Creek Gravels, and it contains Ice Age fossils such as mammoths and peccaries. Just below that lies the 67-million-year-old Denver Formation, which produced the first *Triceratops* ever found. Continue drilling, and you pass through the Arapahoe, Laramie, Fox Hills, Pierre, Fort Hays, Benton, Dakota, Morrison, Lykins, and Lyons formations before reaching the bottom of the hole, where the 300-million-year-old Fountain Formation sits on top of granite. "In this café, you are literally sitting on a big piece of Earth history," I told Ray as he passed me the cream.

Ray's next question was how far down he would have to dig to find a *Stegosaurus*. It was a reasonable question, with a reasonable food analogy readily available. So I ordered a red onion from the kitchen. When it came, I sliced it in half and set one half on the table with the flat side facing up. Formations can be big, but the Earth is a lot bigger, I told Ray, so when the Earth moves, formations can be folded, flipped, or broken. The red onion was my model for the Denver Basin, a geological structure formed by a stack of formations that folded when the Earth moved and the Rocky Mountains were thrust out of the ground. Each concentric segment of the onion represented a formation. In the onion, the segments were horizontal in the center and tilted to nearly vertical at the edge.

The tilted slabs of stone along the front of the Rocky Mountains, such as Red Rocks near Denver or the Flatirons near Boulder, are like the edge of the onion. The horizontal stack of formations beneath the café is in the center of the onion. I put the tip of my fork on the middle of the flat top of the onion. "Here's the Walnut Café," I explained. "In the Rocky Mountains, if you can't see rocks, you're probably on top of them, sitting in a basin."

The answer to the *Stegosaurus* question was simple. Ray could either dig straight down through the café floor to a depth of about 9,000 feet, or we could hop in Big Blue and drive 10 miles to the west, to the edge of the onion, where that layer poked out at the surface at a place called Dinosaur Ridge. We decided that after loading the truck, this would be our first stop.

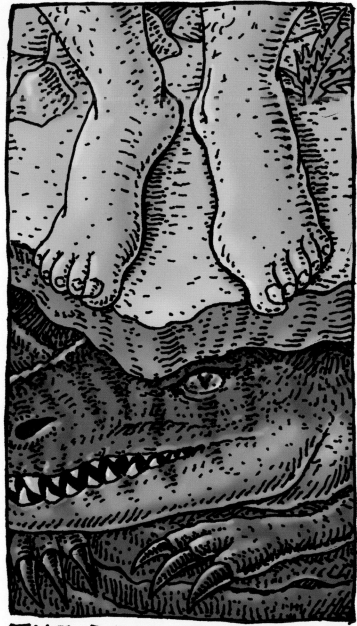

LOOK UNDER YOUR FEET

THE PAST IS THERE

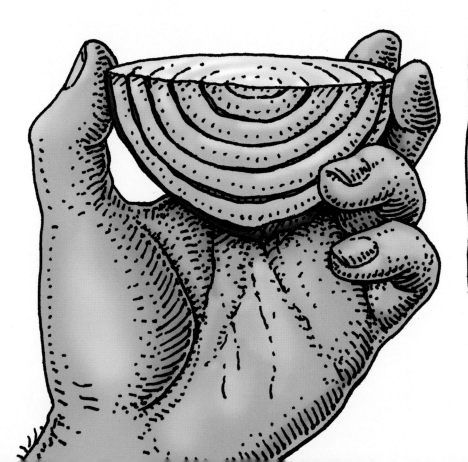

2
ON THE ROAD

No matter how eager you are to leave, it's always a chore to make sure the right stuff gets into the truck. I needed hammers of all sizes, chisels and little picks, my personal pickaxe, sledgehammers, hoes, shovels, brooms, crowbars. Smaller boxes contained superglue, hand lenses, and Ziplocs. To make sure I knew where we were and where we were going, I loaded boxes of geological books, articles, and maps, GPS devices, and my Brunton compass. We packed the toolboxes of Big Blue with cardboard Coke flats, newspaper, tape, Sharpies, and dozens of rolls of restaurant-grade toilet paper (no perforations) to wrap and pack delicate fossils. Then there were lawn chairs, camping gear, propane tanks, coolers, sleeping bags, and the Troll family tent. I filled a couple of five-gallon carboys with water and then went back to my office to think about what I was forgetting. Ray wandered around the museum, looking at skeletons and getting wound up with thoughts of what it was going to be like to find one of his own. I remembered sunscreen, bandannas, a spare hat, and the phone numbers of a bunch of people we hoped to visit. We were as ready as we'd ever be.

Fifteen minutes later we were standing next to Alameda Parkway and Ray had his hands on a *Stegosaurus* bone that was sticking out of a block of sandstone. We were only a mile from where Arthur Lakes had made his big discovery in 1877, and it was only 124 years later. The stone that encased the bone was hard as iron, a condition that by happenstance prevented vandals from chipping the bones out and hauling them away. A diligent group of interested citizens had built a little visitor's center and placed signs along the roadside on the long linear hill aptly named Dinosaur Ridge. This ridge is where the layers of the onion poke out at the surface. The western slope of the ridge exposes the 145-million-year-old Morrison Formation, and the eastern side of the ridge exposes the 105-million-year-old Dakota Group (geologically, a group is a package of associated formations). Since the layers

The famous I-70 roadcut west of Denver exposes the drab Dakota Group and the colorful Morrison Formation.

were tilted to the east and the slope of the eastern side of the hill was tilting at the same angle, the eastern side actually exposes rock that was the surface of the Earth when the Dakota Group was deposited. As we strolled down the road, we could see ripple marks and dinosaur tracks on the tilted surface. By walking only a few hundred yards, we had moved up the stack of pancakes into a younger formation, from *Stegosaurus* bones to *Iguanodon* footprints.

Layered rocks are also known as sedimentary rocks because they began their history as unconsolidated sediment on a landscape. If the sediment is buried by more sediment (the next pancake), and if the weight of the overlying sediment is great enough, the lower layers can be compressed and cemented into stone. Later movements of the Earth can fold this stone and bring it toward the surface of the Earth, where erosion exposes it to the sky.

As we walked along, I pointed out other features in the cemented sand. We could see places where little sea creatures had burrowed into the ripples, leaving collapsed tunnels. In other spots, dents made by driftwood marred the sandy surface. Near the bottom of the hill and the top of the formation, we found layers of black shale that contained easily distinguishable fish bones and scales. It was becoming obvious to Ray that the features of the rock and the embedded fossils were clues about the nature of

this fossilized landscape. He realized, with some gentle prompting, that we were walking on a fossilized beach.

As Ray and I stood looking at our fossilized beach, I explained that depositional areas tend to be low-lying places, as Dinosaur Ridge once was, where water is standing or flowing slowly: rivers, lakes, ponds, swamps, and shallow seas. Sediment settles differently in different environments, and you can see this in the resulting sedimentary rock. The landscape is the canvas on which is painted an ever-changing panorama of ecosystems composed of an ever-changing parade of fantastic plants and animals. Subsequent landscapes bury earlier ones, and some of these beasts and broccoli become fossils.

The Dakota Group sandstone before us was once a beach at the side of a salty sea. But then the setting changed: the region subsided and the sea flooded the old beach, covering it with a layer of gray mud nearly 5,000 feet thick. This layer, known as the Pierre Shale, is itself covered by stark white sandstone known as the Fox Hills Formation, which deposited itself as the sea withdrew and beaches migrated across the top of the mile-thick pile of mud. The whole process only took about 20 million years, and each of those formations are chockablock full of the corpses of animals that died during that time and were buried in the mud and sand.

BEASTS AND BROCCOLI

TWO CONDITIONS ON OUR PLANET

EROSION   DEPOSITION

E WORLD   D WORLD

STUFF DISAPPEARS   STUFF PILES UP

It's a Very Simple Planet

I like to think that the world is composed of only two kinds of places: places where the land is eroding away and places where sediment is piling up. The first is Erosion-World (or E-World) and the second is Deposition-World (or D-World). Fossils form when corpses are buried in D-World; fossils are found when erosion exhumes them in E-World. The perfect place to find fossils is a place that was D-World for much of Earth history and then recently changed to E-World. That is the Rocky Mountain region. For most of the last 500 million years, this region was a flat, subsiding landscape that collected layer after layer of sediments and bodies. Then, about 70 million years ago, the Rocky Mountains themselves were shoved out of the ground, creating local zones of E-World on the mountaintops. The areas between the mountains received the debris that was shed from the rising mountains, and these little pockets, or basins, of D-World accumulated more sediment and more bodies. Finally,

about 10 million years ago, the whole region uplifted, and even the basins became E-World. This last burst of erosion has carved canyons and badlands, exposing 500 million years' worth of fossil-filled rocks. It would be hard to imagine a better setup for finding fossils than the history that shaped this region.

There's a close relationship between erosion and deposition, since the first creates the fodder for the second. Mountaintops suffer erosion, and rivers flowing off mountains carry sand and mud downhill and deposit them in places like New Orleans. The Big Easy should really be called D-Town. I've told my wife that I want to be buried at the mouth of the Mississippi River near New Orleans, because that's the place where I'll have the best chance of getting deeply and quickly buried by sand and mud eroded off the Rockies. It's the best place to actually become a fossil. Conversely, if you get buried in the mountains, you can kiss away your chances of ever being a decent fossil.

Ray, knowing from the interpretive signs that the sea had been a very shallow one, asked how it could accumulate a mile of mud. A good question with an interesting answer. The simple weight of sediment can cause the Earth to flex down. This flexing makes room for more sediment. That's how a 600-foot-deep sea can accumulate 5,000 feet of mud. It's like those spring-loaded plate-dispensing devices you see at cafeterias. You take a plate, and another rises into its place. When you load those things, you stick in plates one at a time, and the top of the stack stays in the same place because the increased weight of the stack of plates is pushing the whole thing down. In the same way, sediment layers can thicken, one formation can bury another, and the accumulating weight of sediments can cause the Earth to flex down. Ultimately, this is how the rock column came to be.

Ray's head was about to explode. He had come for fossils, and he was getting rocks and a long discourse on geology. But I had to make one more point. Because specific formations contain specific fossils, and because formations come in a predictable sequence, all you must do to find the fossils you want is to pay attention to the formations. That way you know where you are in the geological column and, thus, where you are in geologic time.

Whenever I drive anywhere, I'm aware of the formation that I'm driving on, even when I can't see it. I do this by knowing where I started and paying attention to small road cuts and the direction the formations are tilting. That way, I know if I'm driving up the section into younger formations or down the section into older formations. Occasionally, I may be lucky enough to see a great outcrop of an obvious formation, such as the bright red Chugwater Formation. This lets me check my bearings. I promised Ray that he could, at any point in the trip, ask me what kind of rock we were driving past and what type of fossil we would find if we stopped the truck. It's not that brash of a claim if you understand geology and know your formations.

Mountains Are Insignificant, Continents Wander, and Seas Come and Go

The world is old, about 4.567 billion years. That's 4,567 million years, or 45.67 million centuries. This is a really big chunk of time. When you have a lot of time to work with, big permanent things don't seem so big or so permanent. Take mountains, for instance. Mountain ranges seem big to us because they're so much bigger than we are, but we should really compare mountain ranges to the size of the globe. Then they don't seem so big. Mount Everest, the tallest mountain today, is 29,058 feet tall, or about six miles. Seems big until you realize that it's located on a continent that is almost 3,000 by 5,000 miles. In fact, if the Earth were the size of a billiard ball, it would be smoother than a billiard ball. So maybe mountains aren't that big. But how permanent are they?

We know that wind, water, and freezing and thawing cause rock to be slowly worn down over time. There are about 25 millimeters in an inch. If we assume that it takes a year to weather down one-quarter of a millimeter (one-hundredth of an inch) of rock, then it only takes 4 million years to wear down a kilometer of rock, or 6.4 million years to get rid of a mile of rock. At this rate, we could rasp Mount Everest off the face the Earth in less than 40 million years. Remember that the Earth is 4,567 million years old, more than enough time to get rid of Mount Everest a hundred times, and you begin to realize that mountain ranges can come and go. As Donovan sang,

"First there is a mountain,
then there is no mountain,
then there is."

That's important to realize, because mountains have come and gone in what we now call the Rocky Mountain region. In fact, this area has only been mountainous a couple of times in the last 400 million years. At other times, it has been as flat as a pancake. How mountains grow is a ques-

FIRST

THERE IS A MOUNTAIN

THEN

THERE IS NO MOUNTAIN

THEN

THERE IS.

tion that is still being worked on. For the moment, let's just say that mountains are pushed up by forces below. Some of these forces are related to plate tectonics, and others seem to be associated with the movement of heat within the Earth. Since mountaintops are E-World, it's often a race between how fast a mountain range is being pushed up and how fast it's being worn down.

The deepest point in all the world's oceans is about a mile and a half deeper than Mount Everest is tall, about 37,000 feet. Most of the world's oceans are much shallower than that, say 10,000 to 12,000 feet. But seas are shallow, on the order of 200 to 600 feet. Seas form when continents flex below sea level and seawater floods the continent. This situation occurs today in the North Sea, the Bering Sea, along the southeast coast of South America, and in many other spots along the edges of the world's continents. It has happened many times in the history of North America. The continent flexes down, and the sea comes ashore. Another way to drown a continent with a sea is for the volume

of the world's oceans to decrease due to volcanic eruptions on the ocean floor. Just like chucking a cinderblock into your bathtub, the water has to go somewhere, so it floods onto the continents. If the volume of the ocean basin increases, then reverse the process and the seas drain off the continent. The final obvious way that sea level changes is by the waxing and waning of the polar ice caps.

So just like that, seas and mountains come and go. I like to say that mountains are nothing and that seas are undependable as well. For that matter, you can't count on continents to stay in one place for long either. For big things, continents move fast. As I write these words, North America is headed west at nearly an inch a year. That may not sound like much, but at that rate, a million years would get North America nearly 16 miles down the road. With 4,567 million years to work with, North America could have lapped the globe three times. That is, if North America had been around that long, which it hasn't. It helps to have a pretty broad view of time to understand this stuff.

3
BONES GALORE

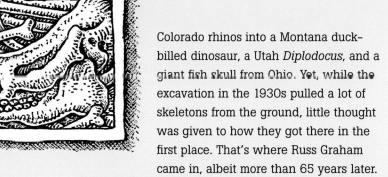

One of the worst things about digging fossils in Denver is driving a big field vehicle in heavy traffic. It was a stinkin' hot Friday afternoon when Ray and I pulled out of the parking lot at Dinosaur Ridge. But the slow stop-and-go start served to remind us that we were headed where people weren't. Fortunately, most roads that head east from Denver don't hold traffic for long, and soon we were sailing toward the northeast corner of Colorado on Highway 76. Our destination was a dig site seductively dubbed Bones Galore.

Though the U.S. Forest Service usually manages forests, somehow it also landed the responsibility for managing some huge swaths of the plains known as the national grasslands. It was on one of these patches that forest service staff noticed a scattering of white fossil bone emerging from the ground in an otherwise nondescript field near New Raymer, Colorado. They contacted the Denver Museum and used the lure of government grant dollars to entice Ice Age specialist Russ Graham to undertake a study of bones of animals that died more than 32 million years before the Ice Age began.

The Denver Museum has a long history of pulling bones from Colorado's upper right-hand corner. During the Depression, the museum used a team of Works Progress Administration (WPA) workers to excavate a quarry and retrieve the skeletons of 34-million-year-old rhinoceroses. Maybe it was the abundant free labor, maybe it was not knowing when to stop, or maybe it was good science, but between 1931 and 1933, the museum quarried the skeletons of nearly 70 rhinos. In the end, it was a good thing, since surplus skeletons make for good trading stock. The Denver Museum was able to parlay its plethora of

Colorado rhinos into a Montana duck-billed dinosaur, a Utah *Diplodocus*, and a giant fish skull from Ohio. Yet, while the excavation in the 1930s pulled a lot of skeletons from the ground, little thought was given to how they got there in the first place. That's where Russ Graham came in, albeit more than 65 years later.

Russ Graham has made a career of studying how animals die, get buried, and become fossils. For him, the Bones Galore site was close enough to the museum's old digs to suggest that he might be able to get some new information to perk up the old skeletons. As a forensic paleontologist, Russ was going to treat this fossil site like a crime scene.

After Ray and I had been driving for a few hours, we were passed by a grinning Russ, who was driving another museum truck. Rush hour had given way to the open road, and the person we were looking for had found us. We followed him to a gas station and then the next few miles toward his camp.

As we headed for the dig site, I began describing the landscape to give Ray a context for what we were seeing. The plains of northeastern Colorado are rolling grassy hills occasionally incised by deep, narrow gullies that expose the shallowly buried bedrock. The rock layer beneath this subdued landscape is the White River Group. White River rocks are indeed white, or at least a very light gray, a direct clue to their origin as airborne volcanic ash. Being from Seattle, I know what happens when volcanoes erupt. When Mount Saint Helens blew on May 18, 1980, snotty magma in the volcano's neck exploded into the sky and flash-froze into silt-sized particles of glass. This volcanic ash rained down on the surrounding landscape like sleet that wouldn't melt. Eighty miles east of the mountain, the fruit-growing town of Yakima saw some wicked storm

clouds that turned out to be
made of rock rather than water.
Within hours, the ash was falling
on the town, collapsing roofs
and burying the ground under
eight inches of ash. Ray nodded
as I told him this, remembering
that he was playing softball in
Pullman, Washington, on that
fateful day, and the game was
called because it began to snow
rock dust. Farther east, the
accumulation was thinner, but
the cloud crossed the country
in three days, and the fine dust
settled the whole way. Volcanic
eruptions like this are one way to move mountains: simply
blow up a mountain and let the prevailing winds do the
work.

The volcanic eruptions that formed the White River
Group, the bedrock of northeastern Colorado, were very
similar. The rock is mainly composed of volcanic ash that
blew out of a series of volcanoes and calderas in Utah
and Nevada between 35 and 25 million years ago, fell out
of the sky onto Colorado, and was mixed up and moved
around by rivers. I like to think of this as a way of moving
one state over to the next. It would have been a terrible
time to be in Nevada, what with the state exploding and
all, but it wasn't so nice on the plains either. A sporadic
but regular weather pattern of airborne rock dust repeatedly buried the landscape in thick layers of dusty ash.

In places, the White River Group is several hundred
feet thick. Not all of this rock fell from the sky, but much
of it did. Streams flowing off the Rocky Mountains then
moved and shifted the ash and mixed it with sand and
gravel from the mountains. The resulting sediment was a
composition that was splendid for the fossilization of bone,
and the animals that lived and died on this landscape
were often buried and fossilized. It was Russ's job to
determine just how that happened.

To get to the camp and fossil site, we drove past

the town of Raymer and over a hill, spotting a mule deer
with such a large rack, it looked like one of those cutout
silhouettes that are so popular on the hills outside western
towns. The road to the site was a two-track across a bleak
field, and the site itself was seemingly unspectacular, a
colorless trench at the base of a low hill. The sun was
just setting as we walked around the site and saw a few
fragments of white bone exposed at the bottom of a hole.
Russ's team had recovered bones from the *Megacerops*,
a knob-nosed rhino-sized titanothere; *Subhyracodon,* a
flat-faced horse-sized rhino; *Mesohippus*, a graceful
deer-sized horse; and *Leptomeryx,* a rabbit-sized deer
relative. Noticeably absent were bones of *Trigonias*, the
museum's eminently tradable rhinoceros. All in all, this
was a typical assemblage from the Eocene portion of the
White River Group.

Many of the bones were crushed, and tooth marks
suggested that something had gnawed the carcasses. Russ
had begun to suspect that all this chewing was the work
of the giant piglike *Archaeotherium*, an animal that would
haunt us for the rest of the trip.

The failing sun drove us to camp, where Russ held
court to a team of 20 Denver Museum volunteers. The
museum runs a program to train adult amateurs in the
ways of paleontology, and the graduates are fossil fanatics

of the first degree. This field course was the culminating session of the program, and the volunteers were full of energy and eager to meet the Troll. They had an upper quiet camp and a lower rowdy one, where warm beers and wet chocolate bars floated in coolers full of melted ice.

We gravitated to the lower camp, where the bright lights and uncovered food dishes had conspired to create a nasty spread called Bug Butter. But the talk was about fossils, not culinary delights. Russ had recently uncovered a rhino skeleton that was essentially intact, except that it was missing a panel of ribs. From his work in the Yukon, he knew that wolves often access the tasty guts of downed caribou by cutting out panels of ribs, and he was pretty sure that Eocene pigs had been up to something similar. Ray was fascinated with potentially carnivorous scavenging pigs. His eyes grew wide, and I could see his sketching hand starting to quiver. We fell asleep dreaming of being gutted by prehistoric pigs and woke the next morning to the vengeful smell of frying bacon.

HELL PIG

ARCHAEOTHERIUM

In better light and with a full stomach, the quarry scene made more sense than it had the night before. Russ and his team were methodically gridding off the area and treating it like an archaeological site, where the position of every bone is carefully recorded before excavation. Using these techniques, Russ was able to reconstruct the field of bones as it was before it was buried and come to some conclusions about how the carnage had unfolded. He had become convinced that the site was the remains of an old water hole and the bones were those of animals that died of starvation. Water holes are the places of last refuge during drought, but they can be death traps, since the increasing crowd of animals eats all the nearby foliage and is trapped in the untenable position of having water but no food. And then the water dries up and the animals die.

As morning wore on, the eastern Colorado sun became hot and the white rock made the landscape unbearably bright. I began to feel like we, too, were at the edge of a diminishing water hole. As much as I liked the concept of Eocene rhinos and giant killer pigs, it was time to hit the road.

THE STUFF THAT HAS ERODED AWAY

WHAT USED TO BE...

THE LAYERS BELOW

LOOKING AT OUTCROPS

With our brains full of thoughts of death, thirst, and starvation, we drove north toward Nebraska, passing the paired Pawnee Buttes. I pointed out the window at them, ready to resume my landscape lesson for Ray.

Buttes are like landlocked icebergs, surface expressions of much more below. In places where erosion has planed the landscape flat, it's difficult to know what lies beneath; the only outcrops are deftly hidden in gullies or streambeds. The Pawnee Buttes stand high above the landscape and are surrounded by nothing. And in geology, the presence of nothing often tells a story.

As I told Ray, I prefer the three-step method for thinking about an outcrop. Step one is to look at the outcrop and try to understand it. Step two is to project it underground and try to understand where it goes. This allows you to predict where you can find it again. Step three is to project the layers of the outcrop into the air and think about how much has been removed by erosion. Often the space where the outcrops used to be is a much better story than the outcrop that is still there. The Pawnee Buttes standing high above the plains meant that the whole area had once been covered with a thick sheet of bone-bearing volcanic ash. The buttes were whittled out of this slab of ash, and their presence is evidence of its very existence. Ray likened this way of seeing the landscape to the way artists see the negative space around a subject as an aid to visualizing the subject.

The upper portions of the Pawnee Buttes are composed of Oligocene and Miocene rocks that were formed as streams poured off the Rockies and scoured up the underlying layers of the White River Group. Farther to the east, in Nebraska and South Dakota, these layers lie at the surface in the form of badlands, gullies, and banks that have yielded some of the best fossil mammal sites on the planet. With names such as Brule, Arikaree, and Valentine, these formations preserve an insanely complete record of mammal evolution from 34 million years ago to the end of the Ice Age, just 11,000 years ago. These were some of the earliest fossils discovered out West; by the 1850s, Joseph Leidy of Philadelphia was describing them. Leidy, a medical doctor, was one of the first scientists to realize the significance of the fossil treasure trove of the American West. He was the first of many, and the White River bones are some of the most thoroughly collected fossils. Museums and careers were made with the discoveries of these old bodies.

As we headed north, we skimmed the edge of Nebraska. I regaled Ray with tales of the treasures found there. One great site is Agate Fossil Beds National Monument in Nebraska's panhandle, located on the site of James Cook's ranch. Cook was a cowpuncher and Indian scout who met the paleontologist O. C. Marsh in 1875 at Fort Robinson and grew interested in fossils himself. A few years later, he met Kate Graham, the daughter of a Cheyenne doctor, and while they were dating, they found some bones on her father's land. A year after they got married, James and Kate moved onto the land, named it the Agate Springs Ranch, and, in 1887, had a kid named Harold. It wasn't until Harold was 17 that paleontologists descended on the ranch to explore for fossils. What they found was an amazing 20-million-year-old bone bonanza. The central find was a continuous layer of little two-horned rhinos called *Menoceras*. This layer was so dense that several museums showed up and chiseled out square chunks, just like cutting brownies out of a pan. The "brownies" were up to 10 feet on a side, and everybody wanted one. Teams from Princeton and the Carnegie Museum were quickly followed by scientists from Amherst College, the American Museum, Harvard, the Denver Museum, the University of Nebraska, and more.

Common Names, Scientific Names, and Rampant Confusion

One of the challenges of talking about fossil plants and animals is what to call them. People have common names for living animals and plants they know such as deer, antelope, cantaloupe, beaver, and pear, for example. Scientists point out that even with living organisms, common names cause confusion because they can be applied by different people to different species. My bear may be a grizzly while yours is a black. More than 300 years ago, a Swedish botanist named Carl Linnaeus solved this problem by developing a system to apply scientific names to all living organisms. The Linnaean system assigns a two-word Latin name to each species. The first word, known as the *genus*, is analogous to your last or family name, while the second word, known as the *species*, is like your first name. For a new kind of organism to get a name, a single specimen is chosen as the name holder, or type specimen, for the species. That means that for every organism with a scientific name, there's a single, "chosen" example of that organism in a museum somewhere in the world.

When I was a kid in Seattle, we had a big-leafed maple tree in the backyard. I called it a maple tree. Years later, I learned that this species of tree was discovered by Lewis and Clark on the banks of the Columbia River near Astoria, Oregon, in 1806. They collected a flattened branch and leaves and brought them back to Philadelphia, where a botanist formally named the species *Acer macrophyllum* (Latin for "big-leafed maple tree"). To this day, you can go to the Academy of Natural Sciences in Philadelphia and see the branch that Lewis and Clark collected. What's not so obvious is that all living examples of *Acer macrophyllum* are only considered to be in that species if they are truly like the type specimen in Philadelphia.

The problem with fossilized plants and animals, of course, is that they're long dead, so people weren't around to give them common names. We often end up using scientific names for common names: *Tyrannosaurus rex*, for example. This partially explains why some scientists sound like they're speaking Latin: because they are. It also explains why some people fall asleep when scientists are talking. To counter the problem of sleeping people, some paleontologists make up common names for extinct animals or for groups of related extinct animals. *Apatosaurus ajax* is a type of dinosaur known as a sauropod, but some people find it easier to simply call them long necks. For other people, the term *long necks* brings up another image.

Throughout this book, we'll use the genus (you can tell because it will be capitalized and italicized) or the genus and species (two words, both italicized). We'll use common names if they exist and descriptions if they don't. Terms such as prongbuck and bear dog are common names used to describe extinct genera.

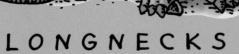

LONGNECKS

Mary Dawson, the Carnegie Museum's Arctic-exploring paleomammalogist, with the original skeleton of *Daeodon hollandi*.

Young Harold Cook had his education delivered to him, and he couldn't help but become a paleontologist himself. Other fossils from the site included *Stenomylus*, a darling little greyhound-sized camel, and *Moropus*, a claw-footed, horse-headed chimera of a beast. Agate Springs's pièce de résistance was a giant skeleton discovered by a team from Pittsburgh. And what a skeleton. *Daeodon* (formerly *Dinohyus*) *hollandi* was a massive entelodont, more than twice the size of the piglike beasts that Russ blamed for ripping the ribs out of his chomped rhinos back at Bones Galore. The beast was cruelly named for the Carnegie's controversial director W. J. Holland, and Pittsburgh's papers had a field day with headlines blaring "*Dinohyus hollandi*, The World's biggest Hog!" and "Porkers weighed two tons." Its front teeth look like broken baseball bats, and the whole body is bigger than a bull bison. Here was a predator that could eat rhinos.

But *Daeodon* wasn't the only terror at Agate. Recent excavations have discovered big bear dogs buried in their dens. When I mentioned them, Ray, who had recently been to New York to visit the American Museum's huge display of fossil skeletons, recalled a particularly memorable skeletal vignette of a voracious grizzly-sized bear dog (*Amphicyon*) in hot pursuit of a fragile and asymmetrically antlered prongbuck (*Ramoceros*) from the 14-million-year-old rocks of northeastern

Colorado. The little prongbuck looked like an ancestral jackalope, but the bear dog was no joke.

Rolling through Nebraska, Ray and I started chatting about the fanatical devotion to fossil collecting that grips some people, and that made me think of paleontologist Morris Skinner, a Nebraska kid who devoted his life to fossils. He spent his career working for a Long Island philanthropist named Childs Frick, who had the cash and the inclination to bankroll serious fossil collecting. Skinner, who spent nearly 60 years amassing the world's largest collection of fossil mammals, once said, "I am one of those people who just have to hunt fossils." The Childs Frick collection, donated to the American Museum of Natural History in the 1967, consisted of 7,400 crates of fossils and $7.5 million in cash and was the largest gift the museum had ever received.

The collection now occupies its own seven-story building attached to the massive museum complex on Central Park West. Skinner had a particular fondness for fossil horses, but he didn't let that stop him from filling the building with everything from rhinos to rabbits. Visiting the collection is like going to a giant department store for extinct animals. Floor one has the killer pigs, oreodonts, and deer; floor two is reserved for elephants; floor three

A *Moropus* and a boy.

TUSKER TIPPING

has herds of horses; floor four is all about pronghorns and bison; floor five is marsupials, sloths, rodents, carnivores, whales, and a host of other oddballs; floor six is full of rhinos, titanotheres, and chalicotheres; and floor seven is for camels.

Collecting mammals gets serious when elephants show up on the scene, and in Nebraska, Skinner and

others have found an elephantine Valhalla. Fourteen million years ago, ancestral elephants in the form of a group known as gomphotheres wandered across the Bering Land Bridge and into North America. Some of these big elephants had tusks on both their lower and upper jaws. And some, such as *Amebelodon*, had lower tusks that were flattened like shovel blades. What these shovel-tuskers did for a living is hard to say. In 1962, two workers laying an electrical line near Crawford, Nebraska, found a big bone sticking out of a bank. They took a chunk to nearby Fort Robinson, where a University of Nebraska paleontology crew was staying. The crew started to dig the site, and identified the animal as a Columbian mammoth. They were initially confused when the skull appeared to have four tusks. Then they realized that they had found one of the most amazing fossils of all time. Instead of one mammoth, they had two bull mammoths that had died with their tusks locked together in a true fight to the finish.

Nebraska has 93 counties, and 90 of them have yielded elephant skeletons or bones. In fact, western Nebraska has so many fossil mammals that the state university has held contracts with the state highway department to salvage prehistoric skeletons that lie in the way of roads. The state museum in Lincoln has one of the finer displays of fossil mammals in the country. It features 11 different fossil elephants and an amazing 11-foot-tall camel known as *Gigantocamelus*. This humped giant weighed more than twice as much as any living camel, and it came from a 1936 discovery at a place called Lisco in Garden County. That same site produced bones from more than 70 individual skeletons: a giant herd of giant camels.

Another site in Antelope County preserves a herd of Pliocene rhinos (*Aphelops*), buried under an ashfall. The animals are so well preserved that one of the pregnant females has a full-term rhino fetus preserved in her rib cage. All this from a state whose population chooses to call themselves cornhuskers.

Ray had dreams of seeing the rhinos of the Ashfall site and the hall of elephants in Lincoln, but our destiny lay to the west, where we had a party to attend, so we rolled onto Interstate 80 and headed into Wyoming.

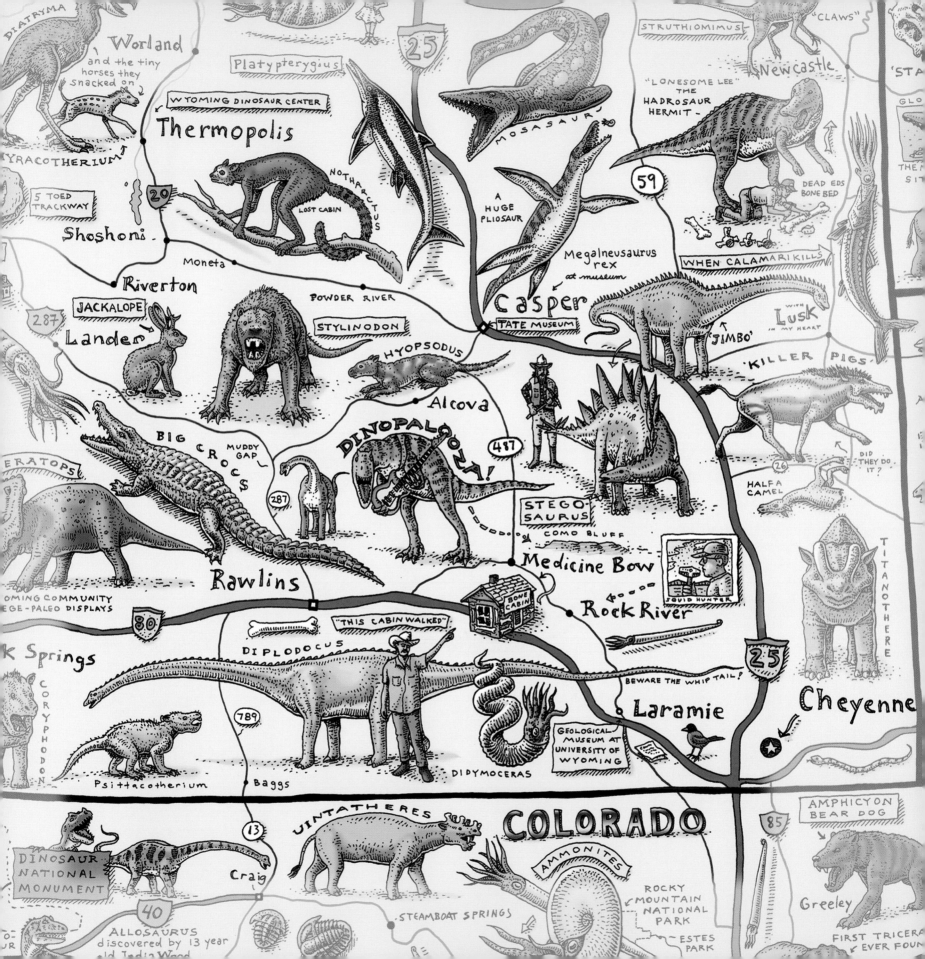

4
THIS CABIN WALKED

For me, when traveling in Wyoming, a stop at Little America is mandatory. What is for most people an elaborate truck stop–hotel–golf course combo is for me a family moment. My mom, who grew up in Wyoming, used to tell me about a sheepherder who was trapped by a blizzard west of Rock Springs in the 1890s. He prayed that if he survived, he would build an oasis for travelers on that very spot. He survived, and the original oasis opened in 1934. Now there is a chain of half a dozen overbuilt compounds where survival is no longer an issue. These are truck stops for the common man, with phones at the tables and sweet waitresses who call you "sugar" and serve up tuna sandwiches and ice cream. Whenever I'm at Little America, I honor my Wyoming survivalist heritage and call my mom to let her know that I'm okay. Ray is always good for a cup of coffee, so we grabbed a table and I had a nice chat with Mom. Then we headed up a long hill known as the Gangplank.

West of Cheyenne is the only place where erosion has not severed the connection between the Great Plains and the Rocky Mountains. The Rocky Mountains were built in several steps. First the core of each range was pushed up, folding the overlying layers and sending sheets of sediment out into the adjacent basins, partially filling them. Some of the basins filled to the brim and buried the ranges in their own debris. Then the whole region lifted, and erosive rivers cut deep canyons into the ranges and basins, carrying the debris out of the region and off to the drainage of the distant Mississippi River. This process scoured out the basins more rapidly than the more resistant mountains, and the mountains began to grow, this time more by the lowering of the basin floor than by the raising of the mountains. In many places in the Rockies, you can see smooth, sloping surfaces that ramp up toward the mountains. These ramps are the remains of the surfaces that existed when the basins were filled. The subsequent erosion has dissected the sloped surfaces, making tilted buttes. The result is a series of ramps that slope toward the mountains but never get there. The Gangplank is the best exception to this process. It's a place where the basin-filling slope is still intact.

Driving west on Interstate 80, which follows the Union Pacific railroad tracks, which follow the Oregon Trail, we were taking the simplest route to cross the Rockies. When we reached the top of the Laramie Range, we stopped at the Vedauwoo Campground and I unrolled the geologic map of Wyoming to continue Ray's rolling Geology 101 class. It was important to me that my artist friend comprehend the geometry of geology.

Geologic maps look like you would need an artist to interpret them. Swirls and splotches of color overlay the familiar landscape. The patterns are the result of history, the culmination of D-World and E-World repeated again and again, with a bunch of other stuff mixed in.

The critical insight to understanding a geologic map is that each color is keyed to a distinct group of rocks exposed at the surface. Sometimes these rocks are the result of deposition and are layered. Other times the rocks are cooled lava that splatted out of a volcano, and they lie on the landscape like seagull poop on a pier. Faults break and move rocks; forces fold and deform rocks; new sediments bury old rocks; and recent erosion slices away overlying rocks to expose older ones. All these processes live on the colored surface of a two-dimensional geologic map. Wyoming's was compiled by David Love, the geologist who is the central figure in John McPhee's book *Rising from the Plains.* The title of the book is a double entendre. The Rockies appear to rise from the plains and they literally rose from the plains. Since the individual ranges of the Rockies are oriented in a variety of directions and because they pushed up through thick overlaying piles of sedimentary rocks, Dave Love likened the formation of the Rockies to so many pigs waking up under a blanket.

Dave published the Geological Map of Wyoming in 1955 and then he did it again in 1985. It's a phenomenal piece of work, and, with it, any trip to Wyoming is so much better because every hill, road cut, or distant vista starts to make sense. The details of the mountains, the basins, and the geologic formations are on the same sheet as the roads, towns, and rivers. The map really is a paleontological treasure map. But after a while, I could see that Ray's interest was waning, so I rolled up the map and we rolled down the vast slope of the Laramie Range and into Laramie, a town locked in time.

The University of Wyoming is a venerable institution, founded in 1886. My Uncle Leroy was a vaunted running back for the Wyoming Cowboys back in 1950, and the place doesn't look like much has happened since then. But that certainly doesn't mean it's not worth visiting. A one-room geology museum on campus has some of the greatest fossils from Wyoming. Laramie is not far from some of the best dinosaur country in the world, and the room is dominated by the skeleton of a monstrously huge *Apatosaurus*, one of the bulkiest of the long-necked dinosaurs to ever flatten a landscape. The museum is a one-man show, run by the extremely blonde Brent Breithaupt, who curates, tours, talks, and performs all the museum functions. If you can get it, a tour of Brent's office provides a great example of how sediment accumulates. His desk was long

ago buried in papers and books, and now all you see is a haystack of printed debris in the middle of the office. Sure enough, in keeping with the principles of basic geology, the youngest layers are on top.

The Denver Museum had borrowed a six-foot fossil garfish from Laramie for the *Cruisin' the Fossil Freeway* exhibit, and Ray had done a splendid job of painting the wooden museum case that held the giant fish. Ray was keen to see how Brent displayed the gar on its home field, and we also wanted to check in on the Laramie *Apatosaurus*, the biggest dinosaur in the smallest museum around. It was not to be. Brent had organized a *T. rex* run the day we arrived, so the museum was shuttered. We paused briefly to stock up on beer and beans and then headed north out of town toward Medicine Bow. We were determined to attend a party in the middle of nowhere, known as Dinopalooza.

About 40 miles out of Laramie, Highway 30 arcs around to the west and a rounded high hill fills the skyline. This is Como Bluff, the first true American dinosaur Shangri-la.

Dinosaur discoveries near Morrison, Colorado, in March 1877 had shown the Morrison Formation to be rich in bones, but the rock was hard as steel and the digging was extremely difficult and slow going. Later that year, a bored Union Pacific railroad employee realized that the mysterious loglike rocks on Como Bluff were actually ancient bones, and he contacted Yale paleontologist O. C. Marsh, who hired the man and immediately sent some others. At Como, the rock was soft and there was little overburden. Bones began to pour out. The spectacular *Apatosaurus* in the great hall of dinosaurs at the Yale Peabody Museum came from this hill, as did dozens of others. The rush for big dinosaurs was on, and it didn't abate until the 1930s.

In 1898, a group from the American Museum of Natural History led by Henry Fairfield Osborn, an imperious American aristocrat, and the impeccably dressed Barnum Brown, was prospecting for bones near Como Bluff. They stopped to ask directions from a lonely sheepherder who lived in a stone house on the desolate windblasted land to the north of Medicine Bow. The sheepherder didn't have any information, but then the bone diggers realized that the stone house was actually a bone house. So numerous were the bones that the sheepherder had inadvertently built his home out of chunks of fossilized dinosaur. The resulting Bone Cabin Quarry was excavated for the next decade and yielded tons of dinosaur bones, including the mighty *Apatosaurus* that was mounted by Osborn's men at the American Museum in 1905. This was the first sauropod dinosaur to be publicly exhibited, and the world took notice.

In November 1898, news of the American Museum discoveries at Bone Cabin Quarry caught the eye of steel magnate Andrew Carnegie. Reading the morning newspaper in his Manhattan mansion, he noticed a headline that proclaimed "Most colossal animal ever on Earth just found out West." He couldn't imagine not owning the biggest dinosaur in the world, so he penned a note to W. J. Holland, the director of the fledgling Carnegie Museum, with the famous line "Buy this for Pittsburgh." The $10,000 check that accompanied the note launched the Carnegie Museum dinosaur program. Holland's field men found a skeleton at Sheep Creek, Wyoming, on July 4, 1899, that they named *Diplodocus carnegii*. Andrew was a proud man, and when the king of England requested a cast of the animal for the British Museum, Carnegie was more than happy to send an example of an American dinosaur bigger and better than anything found in old Europe. He got so much acclaim for this act that soon museums in Paris, Rome, Vienna, St. Petersburg, Madrid, Mexico City, and La Plata sported their own plaster casts of Carnegie's *Diplodocus*.

The *Apatosaurus* at Laramie was collected in 1901 by the Carnegie Museum, repatriated to Wyoming in 1956, and installed in the University of Wyoming museum between 1959 and 1961. Repatriation is a sentiment that still has traction in Wyoming, where "Keep 'em in Wyoming" bumper stickers can be seen decorating the trucks of local bone diggers.

By the time Osborn retired from the American Museum in 1930, the blush was off the rose of American

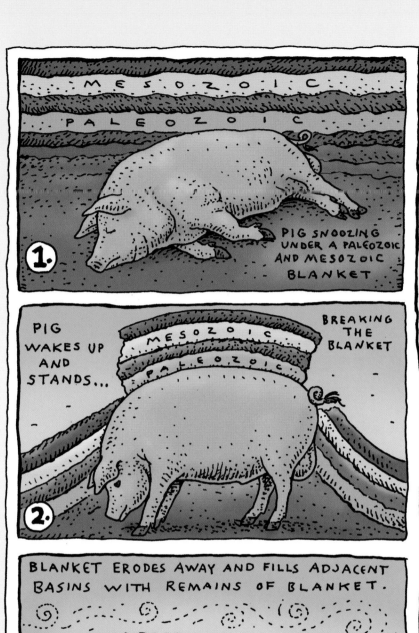

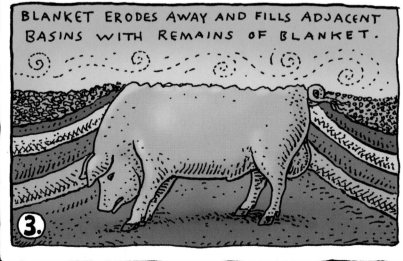

paleontology. The Great Depression depleted field budgets, and other types of science were on the rise. The first big bone rush was over.

Today, Como Bluff is the lair of America's best-known and most notorious paleontologist, Dr. Bob Bakker. Bob was an undergraduate at Yale in 1968 when John Ostrom discovered a startling little dinosaur named *Deinonychus* near Bridger, Montana. The anatomy of the wolflike *Deinonychus* and a careful examination of the few specimens of the half-bird, half-dinosaur from Bavaria known as *Archaeopteryx* led Ostrom to reopen the discussion about the nature of dinosaur metabolism and the relationship between birds and dinosaurs. Ostrom was a meticulous and insightful anatomist. Bakker, clever and charismatic, was a quick study and a splendid artist who penned insightful and convincing drawings of dinosaur anatomy. With Bakker's art and vivid imagination and Ostrom's rock-solid anatomy, the two leveraged the rebirth of dinosaurs. What the Depression, World War II, the Manhattan Project, and the GI Bill had taken away by thoroughly distracting most everyone from paleontology, Bakker and Ostrom gave back. By the 1980s, the Dinosaur Renaissance was underway, and more than just children were paying attention to dinosaurs once again. Bakker's big book, *The Dinosaur Heresies*, published in 1986, is a beautiful, blustery tome that makes paleontology fun and interesting. Ideas from this book found their way into Michael Crichton's *Jurassic Park*, which was a major bestseller in 1990. Bakker started to challenge Stephen J. Gould as the world's best-known paleontologist.

Bob's ideas hit Hollywood in 1993 with the debut of the movie *Jurassic Park*, but Bob wasn't there with them. Steven Spielberg had tapped Montana dinosaur digger Jack Horner, not Bakker, to be the main adviser for the film. In the *Jurassic Park* sequel, *The Lost World* (1997), fatal injury was added to insult when a bearded Bakker-like character was eaten by a *T. rex*.

The real (and uneaten) Bakker always wears his signature outfit: an absolutely battered straw hat, a tan field vest, and a long-sleeved field shirt. The hat is the *Flashdance* sweatshirt of headgear, held together by

tiny filaments in danger of disintegration at the slightest breeze. There can't be just one. He must have a room full of them, carefully crushed and tortured to achieve that casually mangled look. The more Ray and I thought about this, the more we started to wonder what Bakker's closet looks like.

Ray was eager to meet the great man, but there was a place we had to visit first. Walter Boylan, a roadside entrepreneur with an eye to making a quick buck, built a facsimile of the Bone Cabin in 1933 on the road south of Como Bluff. And it sits there today: a small building made almost exclusively of dinosaur bones. The place has definitely seen better days. The dream of easy money now an afterthought, some windows are broken and not a lot of attention has been paid to the place's retail potential. A water-stained card stuck in one of the windows reads "Two hundred million years ago, this cabin walked." We stopped, and for the first time in my memory, there was someone there.

Apparently the owners are looking to sell, but in the meantime, they let a laconic rock-picking cowboy named Mike Lewis run the place. Mike looked every inch the quintessential Wyoming cowboy, except for his pair of sandals. It's a challenge to run a place that used to walk, but Mike had some modest plans. Scattered around the bone cabin and adjacent house was a panoply of premium dinosaur artifacts: rusty metal *Brontosaurus* signs, *Stegosaurus* humerus doorstops, and cannonball concretions from nearby marine layers. Ray kept urging me to make an offer to Mike for the metal *Brontosaurus* sign. It felt like offering to buy the fireplace at Old Faithful Inn, and I just couldn't bring myself to do it. Mike told us of his plans to give the sign a fresh coat of paint, and that didn't sound so good either. We left, trusting the fate of the roadside attraction to natural processes, and drove into Medicine Bow. In 2022, we learned the little museum in Medicine Bow is making plans to move the Bone Cabin to town—or as Ray said, "The cabin that walked is about to get up and drive away."

There are places where time stands still, and there are places where it's losing ground. Medicine Bow, Wyo-

Mike Lewis, caretaker of the Bone Cabin, flanked by dinosaur bones at an adjacent building.

Bob Bakker surrounded by paleopaparazzi at Dinopalooza.

ming, is clearly of the latter group. In 1885, Owen Wister, an East Coast dandy, rode west on the Union Pacific, saw the narrative potential of the place, and penned the first Western novel here. *The Virginian*, first published in 1902, spawned a genre that washed over into television and film and made John Wayne and Louis L'Amour some of the biggest names of my childhood. That wave has passed, but the Virginian Hotel, built in 1911, still stands in Medicine Bow, and it feels like Owen Wister just checked out. It is in this venerable establishment that Bakker often holds court.

Just a few years ago, I discovered a bundle of letters from Bert Pearce, my mother's father, that were postmarked in May 1915 from Medicine Bow. He was writing to his fiancée in Denver in anticipation of their wedding later that year. These letters and other records showed me that my grandfather had been cowboying around Medicine

Bow between 1896 and 1915. He was there when the first great sauropods were being dug and when the mythology of the American West was being codified. I knew my grandfather, and his proximity to the origin story of both dinosaurs and cowboys makes me wonder if I chose my profession or if it chose me.

Evening was approaching and the sky was beginning to darken ominously. Realizing that Dinopalooza wasn't going to be a Woodstock-size event, Ray began prodding, "How about a hotel, comfortable beds, and a big chicken-fried steak?" But I knew what he didn't, or at least thought I did, and we rolled north into the waning day and worsening sky. We had good instructions, and after a few miles of gravel we rolled into Dinopalooza. And there was Bakker, half-crammed into his beaten Toyota, fully garbed in his battered hat, and surrounded by the paleopaparazzi. The arrival of Big Blue caused a slight stir in the small crowd. Ray hopped out and joined the crowd around the Toyota to stand in line to shake hands with the Dinosaur Renaissance man.

Dinopalooza is the brainchild of two guys who are living out their childhood dream of hunting dinosaurs. Chris Weege and Dave Schumde are Denver oilmen, each in their early 40s. Both are lanky, lean men, comfortable and smooth at the Petroleum Club, and, at first glance, not a bit nerdy. Using their ample geologic acumen and the curious American fact that if you own the land, you own

the dinosaurs, Chris and Dave have leased their very own dinosaur quarry. That's right, two boys and their very own dinosaur excavation. Chris literally has an *Allosaurus* in his garage. When I, as a museum guy, asked him what he's going to do with it, he just smiled and said he's still thinking about it.

One thing was sure: Chris and Dave know how to throw a good party in the middle of nowhere. Rock-n-Roll Ray, the childhood dinosaur nerd, was in heaven. Your own private dinosaurs *and* live music. Despite the gathering black clouds and what appeared to be a brewing tornado, the band, including at least one mandolin-toting paleontologist, was setting up. Chris showed us his quarry, from which he has extracted parts of a *Stegosaurus* and an *Allosaurus*. He described the plate-backed *Stegosaurus* as stubbier than usual and the *Allosaurus* as one of the oldest known. We scavenged some hotdogs from a thinly populated grill as the wind started to howl. The band started, the rain started, the wind and rain got really serious, the band stopped. Then the rain stopped and the band came back. But it stayed

Dave Schmude and Chris Weege at Dinopalooza.

cold and wet. Given the state of the roads after the only serious rain of the year, there really was no place to go, so the party went on. Dave's little brother, Doug, played some mean slide guitar, and a weird paleobluegrass riff got going. These guys were singing about fossils. "May the belemnite be unbroken, May the ammonite be exalted." Ray happily swilled some red wine he'd picked up in Denver, swaying gently in the gusts. Then a drunken group of partyers led by Chris and Dave lit a *Brontosaurus*-shaped panel of fireworks and the party continued. Late in the night, Chris and Dave told us about another claim they had staked in northern Wyoming, a site rich in fossil fish where they had found and reburied the body of a Jurassic ichthyosaur. This sounded like many of the lost-treasure stories that fueled my childhood lust for discoveries, but in this case, the guys were able to give us precise GPS coordinates so we could find the site for ourselves. Thinking about warm Jurassic seas, we unfurled the Troll family tent and crashed, hungry, cold, and wet, but strangely happy amongst this remote band of fossil fanciers.

We awoke with medium-size hangovers and huge hungers. We had stocked no real food in Laramie, so we were destined for breakfast back in town at the Virginian. Ray had elicited a sketchy promise from Bakker that he would join us for an omelet, so we rolled up the soggy tent and slid the muddy roads back to Medicine Bow. We spotted Bakker just pulling out as we pulled in, and Ray's chance to dine with Bob turned into the diminishing north end of a southbound Toyota.

Fortunately, we had a plan B and had arranged to meet our friend Gary Staab and his pal Kent Hups in Medicine Bow and go hunting for ammonites by smashing them out of giant concretions. Gary, one of the best paleoartists around, is an old friend of ours, but we had never met Kent. He turned out to be

(right)
A trio of Morrison Formation killers (left to right): *Allosaurus*, *Ceratosaurus*, and the giant claws of a *Torvosaurus*.

42

one of those highly focused and utterly driven finders of fossils. Gary and Kent had driven up from Denver joking about hunting squid, since ammonites are extinct shelled relatives of today's tasty calamari. We got our breakfast and headed out for the tiny town of Rock River, where, among other things, there's a fine mural of a curious cowgirl riding a giant pink *Apatosaurus*.

Bashing concretions is about my favorite thing in the world. You get to wander around with as big a sledge-hammer as you can carry and smash open round balls of case-hardened mudstone that, every once in a while, will have a splendid fossil on board. This gets addicting fast, but it's also exhausting and dangerous. When I was a kid whacking concretions on West Coast beaches, I quickly learned that if you hit a concretion with a glancing blow, you have a pretty good chance of creating a deadly projectile. My shins still bear the scars. And if the concretion breaks, the flying shards can take on shrapnel-like properties that make even the most casual digger really think hard about the idea of safety goggles.

Kent and Gary had found a superb spot where giant nodules were eroding out of a ridgetop. These were basketball-size numbers, and most sledgehammers aren't up to that gauge of rock. I had a 12-pounder that just kept bouncing off the case-hardened concretions. But one of

Gary Staab and Kent Hups, the squid hunters, and a *Placenticeras*.

the things you learn is that rock gets tired. Hit it enough, and it'll break. Kent and Gary were whackos, literally. They just kept whacking on the rocks, and in not too long, we had accumulated some splendid ammonites and inoceramid clams, the shiny, ribbed oysters often found with ammonites. As a museum guy, I know that the most common kind of mistaken identifications are meteorites and dinosaur eggs. It must be once a week that somebody comes to the museum fully convinced that they have found one or the other. Round rocks are mistaken for eggs, and heavy or rusty rocks get called meteorites. Concretions, because of their amazing roundness, sometimes get picked for either. Both the roundness and hardness are artifacts of their formation. Most common in the drab gray shale that formed as mud at the bottom of Cretaceous seas, concretions appear to have formed rapidly, probably the result of a chemical reaction between a dead critter and the surrounding mud. A concretion didn't form around every dead body, nor do all concretions contain fossils, but there seems to be a connection nonetheless. It's possible to find fossils in shale that are crushed flat as cardboard, while those in concretions are perfectly three-dimensional and uncrushed. This observation argues that the rock balls hardened before much weight from overlying mud bore down on them. Recently, some marine geologists at Yale have found concretions forming around recently deceased crabs and clams in Long Island Sound. It's a useful natural process, and it's responsible for a lot of cool fossils.

Ammonites are nothing more than extinct shelled marine animals related to squid, octopi, and nautili, but they have a certain something that makes them very desirable. It's hard to explain our lust for ammonites. Like trilobites, ammonites are icons of extinction. So many kinds lived on Earth, and now none do. Ammonites are exquisite objects, coiled shells imbued with the sublime symmetrical essence of the Golden Mean. Ammonites are the shells that she sold by the seashore, she being Mary Anning of Lyme Regis, an

English coastal town near sea cliffs made up of Jurassic marine shale loaded with gorgeous ammonites. Many of the ammonites from concretions still wear their iridescent and pearly shell. If the shell has broken away, the walls of its chambers trace an incredibly delicate dendritic pattern that, once seen, can never be forgotten.

Not all ammonites are smoothly coiled. Deviant species, known as heteromorphs, come in shapes ranging from the G-shaped *Scaphites* to the gently arching *Baculites* to the twisted saxophone–shaped *Didymoceras*. As we smashed the concretions, we were finding pieces of all of these types.

Hauling smashed concretions is a real chore, and it makes you very selective about the fossils you carry home. Kent had found a nearly perfect *Placenticeras,* a smooth, discus-shaped ammonite, and was happy as a clam. Ray and I had found half a dozen partial heteromorph ammonites and were likewise happy. Kent and Gary had to get back to Denver and Ray and I had to head north, so we parted ways. Like hard-core fishermen who always have to make just one more cast, or blue hairs who can't pull themselves away from the slot machines, none of us wanted to stop collecting ammonites.

Ray was tight on cash, and another night in the soggy Troll family tent wasn't an inviting thought, so I called my cousin Lisa in Casper and asked to use her basement floor. After passing back through Medicine Bow and heading north, we drove by the Freezeout Hills to the northwest and rolled into the broad and barren Shirley Basin. This is big, flat country, where cars come along very rarely and pronghorns are everywhere. The long spine of the Laramie Range makes the east wall of the basin, and the Early Eocene Wind River Formation makes its floor. Here and there are petrified logs, remains of the subtropical forests that used to grow here. For much of the 1950s and '60s, this area saw strip-mining for uranium, and the not-so-subtle scars of reclamation are still visible. Lying above the Wind River Formation are the bright white beds of the White River Group, that familiar Late Eocene fossil paradise we saw in northeast Colorado. Forty miles north of Medicine Bow, the flats of the Shirley Basin end at an

escarpment, the lip of a huge mud-floored hole named after a guy called Bates.

Bates Hole is central to my family's history. Bert Pearce was born on the Salisbury Plain in the south of England in 1879. At the age of 16, he and three of his friends saw an advertisement seeking young men to herd sheep in Wyoming. They took the bait and Bert walked away from England and moved to the United States. By the time he was 28, he was working for the Freeland Livestock Company, which owned a big piece of Bates Hole. As the old family story is told, the owners let Bert in as a partner on credit, certain that he would never prove up.

Cracking giant Milk Duds in search of ammonites.

But since it's my family's story, it ended well, and a Casper doctor came through with a loan that allowed Bert to buy into the ranch in 1908.

The ranch was rough, stretching from the muddy gullies in the floor of the hole to the slot-canyons of the Laramie Range. My mom and her four siblings spent summers at the ranch and winters in town. She lived a childhood full of stories of odd old Basque sheepherders drinking vanilla extract for a cheap high and corralling wild mustangs. Her three brothers were wild ones, lowering themselves by rope into ice caves and racing each other backward in jalopies down dirt roads. Bert sold the ranch in 1948, the year my mom left for California, college, and a new life. When I was a kid, my aunt lived in the giant family house in Casper, and we would drive east from Seattle so we could "go out West." The house on Beech Street was an amazing warren full of treasures from a half-century of ranch life in Wyoming. I used to pore over cigar boxes full of arrowheads, polished pieces of Wyoming jade, and fragments of fossil teeth. Uncle Sid, an insurance salesman and my mother's brother, was great for stories about the old days on the ranch and the stuff they used to find, but he had become a townie, and I could never convince him to take me fossil hunting on the old place. I dreamt of finding my own treasures.

Casper, nonetheless, was a place of discoveries waiting to be made. My aunt introduced me to neighbors who collected rocks and to old codgers who were kind and generous with their collections. On a weekend drive to the top of Casper Mountain when I was five, I wandered away from the picnic and starting poking around a big white boulder. I found a sugar-cube-size chunk of rock with the impression of what I now know to be a brachiopod. At the age of five, I was convinced that I had discovered a fossil rattlesnake tail. It was my first big find.

Uncle Sid really annoyed me by telling me about a fist-size dinosaur tooth that he'd found near Headquarters in the '40s. If he still had it, he'd have given it to me. The part that nicked me was that he had given the tooth to a young geologist, and, worse still, he had forgotten his name. No matter how hard I pressed, Sid simply couldn't remember who got what I came to call "my tooth." It wasn't until 1991, a few years after John McPhee had published *Rising from the Plains*, his sublime book about Wyoming geology and Wyoming's premier geologist, David Love, that the mystery was solved. When I told my mom about this new book, she exclaimed, "That's him, that's the guy who has your tooth!"

I had just moved to Denver at the time and immediately got into my car and drove the 130 miles to Laramie, where I stormed into Dave Love's office, demanding my tooth back. The man had a mind like a steel trap. He remembered my uncle, told me some ribald stories about my aunt, and started hunting around in the collections for the tooth. Curation at the University of Wyoming wasn't as efficient as Dave's memory, so we never found it, but the encounter started a great friendship. As Dave pointed out, the geology of Bates Hole is such that the tooth likely didn't belong to a dinosaur. Bates Hole consists of Late Cretaceous marine rocks overlain by Early Eocene rocks overlain by late Eocene rocks, none of which would contain dinosaurs.

While we visited, Dave regaled me with stories of some old German homesteaders in Bates Hole who claimed to have found lots of "tushes" near their cabin. Dave surmised that they meant "tusks" and dispatched a University of Wyoming paleontologist to check out the site. The "tushes" turned out to be the six-inch-long teeth of the Early Eocene hippolike beast known as *Coryphodon*. From Sid's description and Dave's memory, it was more likely that my tooth fell out of the jaw of a Late Eocene titanothere, one of those great knob-nosed contemporaries of the killer pigs. In a later visit to the present owners of the ranch, I laid the family mystery to rest when I saw a toothy titanothere skull sticking out of the side of a hill.

Ray and I talked about my family as the sun set and Big Blue rolled through Bates Hole. Bates Creek, which drains the giant valley, is a tiny ephemeral creek that hardly seems up to the task of such a prodigious excavation, but Wyoming is full of stark and confusing topography. It's a state with some serious history. We stopped for a cup of coffee and to discuss politics with the present owners of the ranch, then rolled into Casper long after dark.

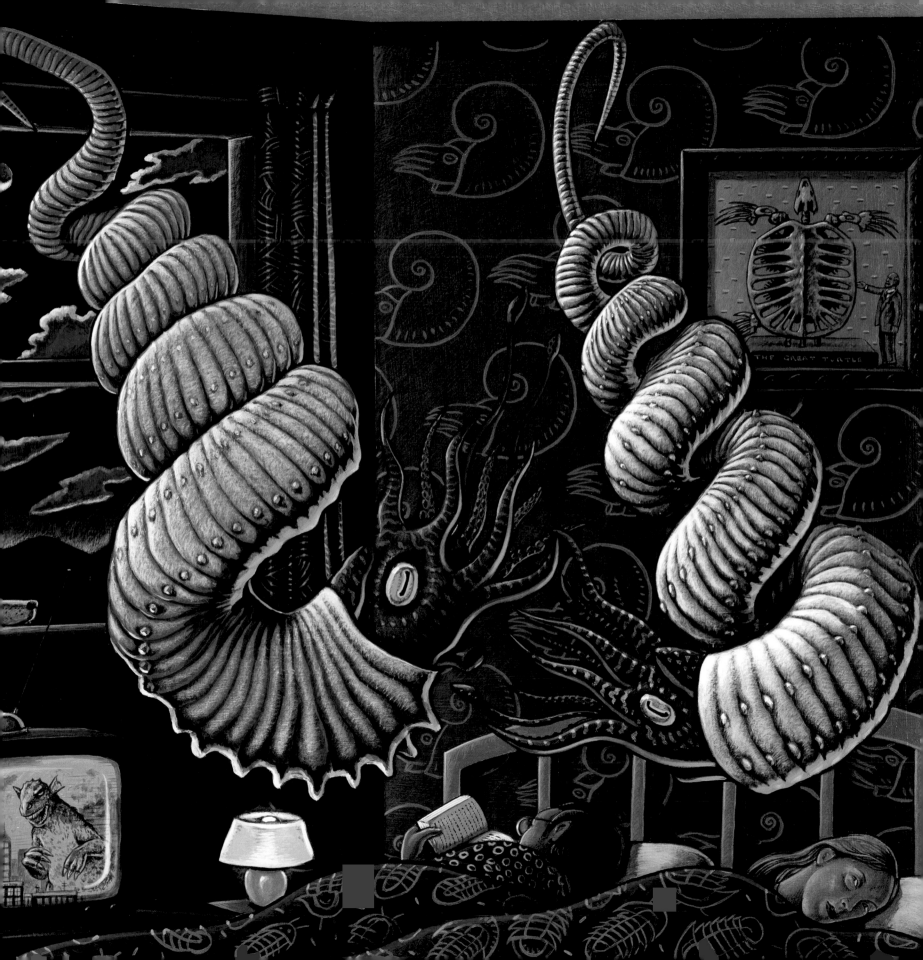

THE GREAT TURTLE

5
THE DEAD EDS

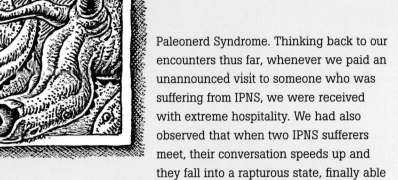

The small city of Casper lies at the foot of Casper Mountain on the banks of the North Platte River. Mildly well-known as a Mormon ferryboat crossing on the Oregon Trail and as an oil and ranch town, Casper has been losing ground since my mom left in 1948. Cousin Lisa and her husband, Stan, better known in Casper as Stereo Stan, have a very skittish border collie and a big basement stuffed with what must be the largest collection of old records and stereo gear in all of Wyoming. Ray and Stan fell into a sort of music-head tail-sniffing thing, each trying to outdo the other by citing their most obscure '60s vinyl treasure. Ray and I ended up sleeping in the basement with the records and the skittish dog.

The Tate Museum at Casper College was our destination for the morning. The few employees and many volunteers of the Tate (known amongst themselves as "taters") are an energetic bunch, and the area is caked with fossils, so the museum's basic plan seems to be to dig fossils, display fossils, and have a symposium every June. Ray and I walked into the exhibit hall and bumped into Bill Wahl and Russell Hawley. Meeting Russell and thinking back to Dinopalooza, Ray and I began to realize that wherever we went, we bumped into people like ourselves: paleonerds. These guys (and most of them are guys) are often lone rangers, one- or two-man dinosaur fan clubs in small towns. It suddenly occurred to us that we had discovered IPNS, or Isolated Paleonerd Syndrome. Thinking back to our encounters thus far, whenever we paid an unannounced visit to someone who was suffering from IPNS, we were received with extreme hospitality. We had also observed that when two IPNS sufferers meet, their conversation speeds up and they fall into a rapturous state, finally able to lapse into paleospeak without fear of being ostracized.

Russell Hawley may be the type specimen of IPNS. He's not from Casper, and he really doesn't fit in Casper. A young, very thin man with black hair, New Yorkish black Goth clothes, and a bright and enthusiastic voice, Russell is there for the fossils. He has an amazing way with a fine-point black pen, and the Tate is festooned with the artistic fruits of his labor. Ray was fascinated. Here was another art-fossil guy, and a good one, gainfully employed as the tater artist-in-residence. Russell binds Xeroxes of his drawings and sells them for next to nothing, with all proceeds going to the Tate. Ray and I tried to buy such classics as *Fossil Reptiles of the Tate Museum Collection*, but Russell would have nothing of it, forcing us to accept our copies as gifts. We happily hauled away stacks of his finely detailed prehistoric scenes. Later, riding out of town in Big Blue, Ray would groan audibly as he flipped through the pages and encountered yet another sweet Hawley image that preempted one of his own prehistoric sketch ideas.

A trio of taters: Bill Wahl, a killer pig, and Russell Hawley.

49

P.N.S.*

* PALEO NERD SYNDROME

THOSE OF US WITH A LIFELONG LOVE OF DINOSAURS

Bill Wahl could not have been more different from Russell. A big, stocky guy like myself, he looked like he'd just been furloughed from the Casper College football team. Here was an artist-scientist pair of paleonerds. These guys were like a different version of Ray and me, in a bizarro world–Superman comic–parallel universe sort of way. Ray and I started to feel a bit self-conscious, but also a bit gleeful as we realized that our road trip idea might have a unifying theme. We weren't just driving around a modern landscape looking at the remains of ancient ones; we were also encountering and engaging people like ourselves who had devoted their lives to the fossil world.

Bill had a big secret: he had excavated a giant dinosaur out of the Jurassic Morrison Formation from a ranch near Orin Junction. This formation shows up at the margins of Laramide basins from New Mexico to Canada, so clued-in diggers can find the big-boy dinosaurs almost anywhere, if they mind their geological p's and q's.

The Morrison is famous for long-necked dinosaurs, and the classics such as *Apatosaurus* (*Brontosaurus* by its other name), *Camarasaurus*, *Haplocanthosaurus*, *Brachiosaurus*, *Barosaurus*, and *Diplodocus* are pretty well-known from many or at least a few decent skeletons. But there's a larger class of beasts known from a couple of partial but huge skeletons: animals such as *Supersaurus*, *Seismosaurus*, and *Amphicoelias*. It looked like Bill had hooked into one of these monsters. He had found a fantastically large, 10-foot-long rib and some spectacularly large vertebrae and called it Jimbo, Tate Museum specimen #21. He was working with Russell and the Tate administration to prepare a massive national press release about the discovery. Standing in the exhibit hall in front of these huge bones speaking in hushed tones and darting their eyes, he and Russell urged us to be quiet about the find. Ray asked Bill why the gigantic specimens were on display, in plain sight, if they wanted them to be a secret. Russell looked at Bill, and Bill looked back at Russell. Neither of them said anything, but it was clear that they weren't too worried about a Tate Museum exhibit accidentally making it into the news.

Although still distracted by Jimbo, Ray noticed a stunningly beautiful killer pig skull in a nearby case and soon was talking a mile a minute with Russell about the killing techniques likely employed by these beasts. As it turned out, the Tate really delivered on this topic. The White River Group outcrops not only in northeastern Colorado and Wyoming's Bates Hole, but also in splendid badlands southeast of Casper, near the town of Douglas. Kent Sundell, a Casper geologist who sells White River skeletons on the side, had recently made a discovery of a group of Oligocene gazelle camels that look like they were mauled by a pack of killer pigs. The specimen, which he sold to the Wyoming Dinosaur Center in Thermopolis, was a big slab of stone with the sheared carcasses of seven camels piled together. Russell had made a pretty drawing of this gruesome scene, which he printed on a T-shirt with the slogan "Attack of the Giant Killer Pigs." We spent some time playing with the giant fossil pig skull, seeing how wide its mouth could open. We were left with the opinion that a killer pig probably could easily slice a little camel in half.

The Tate has some really enticing fragments of other beasts. We inspected marine crocodiles from the Cretaceous Cody Shale, *Didymoceras* ammonites from Rock River, and a tiny little marine reptile known as a nothosaur from the Triassic Alcova Limestone. One specimen that caught Ray's eye was a huge sternum with two enormous swimming paddles. This was a fragment of a much larger marine reptile, a pliosaur named *Megalneusaurus rex* that would have been as long as two killer whales placed end-to-end. The specimen had been found in the Jurassic Sundance Shale in Fremont County in 1898, and the taters were still looking for the rest of it.

Eventually, we tore ourselves away from Russell and Bill and headed east to the town of Glenrock, still fast on the fading tracks of Bob Bakker. Fifty years ago, Glenrock was a small town in the shadow of Casper. Now that Casper has grown into a sizable city, Glenrock is one of those small Wyoming towns that somehow manages to keep on keeping on. We went there because we'd heard that Bakker had set up a museum at a local elementary school. Sure enough, we soon located the Robert T. Bakker Educational Center in the second-grade classroom at Glenrock. The place was shuttered and there was no sign of Bakker or anybody else, so, disappointed, we got on the interstate and headed south.

We stopped in Douglas because we'd heard that it was the official home of the jackalope: the governor of Wyoming deemed it so in 1984. We found a one-horned, six-foot-tall sad sack of a beast on the roof of the mainly closed Le Bonte Hotel and a larger example downtown in Jackalope Square. The badlands south of Douglas are full of Oligocene mammal fossils, including a smattering of fossil rabbits and little deer-like things. Miocene rocks in Colorado and

Kent Sundell and a White River oreodont.

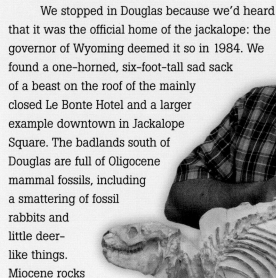

PALEO DEER HUNTER'S DEN

Nebraska have yielded a host of smallish deer with rabbity noses and asymmetrical antlers, known as prongbucks. It occurred to me that the jackalope really wasn't so mythical after all. And it was really amusing to think of all the energy that Wyoming has put into this celebration of a mythical animal when similar real beasts lie buried in the tan rock beneath the eastern half of the state.

Our next stop was Lusk, an old ranching town on the eastern border of Wyoming. It was not too far north of here that John Bell Hatcher found the first good *Triceratops*

skulls in 1888. Hatcher was a gifted field man first associated with O. C. Marsh at Yale, then with a productive but short career at Princeton.

Returning from a trip to Montana, Hatcher had heard some rumors from around Lusk and stopped in on his way south. The story was that a cowboy had found a *Triceratops* skull weathering out of a gully and had roped the horn to check it out. In one rendering, the cowboy brought the whole skull crashing down, riding home with the horn. In another version, he simply snapped the horn off and later took Hatcher back for the skull.

This all played out near the town of Lance Creek, north of Lusk. Here, the Lance Formation of latest Cretaceous age underlies the rolling hills, and the steep-

walled gullies are cut directly into Cretaceous bedrock. Hatcher hit pay dirt, finding hundreds of partial *Triceratops* and collecting dozens, many of which are now on display in the Great Hall of the Peabody Museum in New Haven, Connecticut. One, notably, is at the Smithsonian, where the nation's *T. rex* is munching on a *Triceratops* named Hatcher. *Triceratops* must have been a delicious animal, because its fossil skulls are far more common than its skeletons.

Lance Creek is also where Hatcher figured out how to find fossil mammal teeth by looking at anthills. In the 1880s, there was little knowledge of the mammals that lived with the dinosaurs. Mesozoic mammals were small—raccoon-sized at the largest—and their teeth, the most

Why Leaves?

When I tell people I'm a paleontologist, they usually respond by asking if I study dinosaurs. Then they seem a little disappointed when I tell them that leaves are my preferred fossils and paleobotany is my specialty. After all, aren't plants boring?

The answer is a resounding no. Take a hike and chances are that you won't see any big wild animals, but you will see lots of plants. In most landscapes, it's the plants that make places look like they do, and the total biomass of plants vastly outweighs that of animals. It's always been that way, which means that you're usually much more likely to find fossil plants than fossil animals. This means that I'm often finding fossils while my dinosaur friends are merely hunting for them. It's like the difference between fishing and catching fish. In a typical fossil leaf quarry, my crew and I can easily collect 500 fossils in a single day. More than one of my interns has likened digging leaves to opening Christmas presents. Spend some time in one of our quarries, and you'll hear a steady stream of "oohs" and "ahs" and see a bunch of happy campers.

Add to the instant gratification and fun of finding fossil leaves the facts that fossil plants are beautiful, they carry a signal of ancient climates, and they show what ancient landscapes looked like, and you can easily see why I'm a card-carrying leaf digger.

fossilizable bits, ranged in size from pinheads to peas. Marsh was keen for Hatcher to find mammals, and Hatcher discovered that the mound-building ants sometimes made use of tiny fossils to armor their anthills. Hatcher's insight is still used by mammal paleontologists today: you'll rarely see them walk by an anthill without dropping to their knees and pressing their faces down to the ground in search of fragments of former animals.

I've always been fond of the area around Lance Creek because that's where, in the 1930s, Erling Dorf of Princeton University did the first good studies of the vegetation of the Late Cretaceous. Erling was one of the first paleobotanists to recognize that the plants of the end of the Cretaceous suffered major extinction, just like the dinosaurs did. Erling had been good to me when I was a college student, giving me the keys to the Princeton University fossil collections so that I could prowl around them for long hours late at night. It was an amazing experience to pull out drawers, look at fossils, and talk to the ghosts of William Berryman Scott, one of the first great paleomammalogists, and John Bell Hatcher. Years later, I was able to repay the favor and his kindness by naming a Late Cretaceous plant, *Erlingdorfia*, in his honor.

The famous Sternberg brothers from Kansas also made two spectacular discoveries here, in 1908 and 1911. They found a pair of mummified duckbilled dinosaurs. Somehow the dinosaurs had died and dried before they were buried. The bone fossilized, and the skin was imprinted on the fine sandy mud that buried the bodies. Rather than hacking through the skin impressions, the Sternbergs recognized what they had and were able to extract both skeletons essentially intact. One now resides at the American Museum of Natural History and the other is in Frankfurt, Germany, at the Senckenberg Museum.

By the 1980s, this area was largely forgotten by paleontologists, but the dinosaurs continued to weather out of the soft bedrock. Ranchers noticed the bones and alerted commercial diggers, who came to make a buck. One of these guys, a long-limbed redhead named Lee Campbell, had sold a nearly complete ankylosaur to the Hayashibara Museum in Okayama, Japan. I had seen the skeleton laid out in the backyard of a Tucson hotel in 1991. Today it resides in the town of Fukui, Japan. Since Ray and I were in Lusk, I thought it would be interesting to see what Lee was up to. I had a phone number, so we rang Lee from a pay phone. He started talking about barbecuing steak, and Ray, the carnivore, started to salivate. Lee gave us directions to his house, and we drove on over.

It was about 5 P.M. when we pulled up to the small apartment complex where Lee lived. After greeting us, Lee's first question was, "Do you want to go to church?" Ray said, "Huh?" Lee asked us again if we wanted to go to church, this time adding that it was the place where he stored his dinosaurs. Because Ray's friends the Bonners of western Kansas are fossil diggers who have converted an old stone church into a fossil shop, we were not strangers to the concept. Lee led us to a mildly dilapidated white wooden church. The pews had been removed and replaced with large shelves that sagged beneath the weight of dinosaur bones. The choir area was full of plaster jackets. Workbenches laden with partially cleaned bones lined the walls. Dinosaur skeletons were the only active members of this congregation.

Lee was working a bone bed of duckbilled *Edmontosaurus* dinosaurs. It didn't take Ray long to start the wordplay and shorten *Edmontosaurus* to Ed, and there we were, standing in a church surrounded by the dead Eds. We asked Lee how he'd ended up in this situation, and he told us that he was being bankrolled by a Kentucky dentist who loved digging fossils more than cleaning teeth. Apparently the dentist paid the bills, and Lee dug up the dead Eds and stored them in the church. It seemed like a reasonable arrangement, in a Lusk sort of way.

Since the hot sun was still high in the sky, we decided to postpone the steak, much to Ray's dismay, and take a drive out to the dead Ed quarry. Lee led and Ray and I followed in Big Blue. As we headed north from Lusk, the Black Hills became visible in the northeast and the rolling hills gave way to the flat valley of Crazy Woman Creek.

We passed Redbird, Wyoming, a famous place if you care about the Cretaceous Interior Seaway. Between 90 and 70 million years ago, most of the Western Interior was covered by a shallow, salty sea. Mud accumulated at the bottom of this sea, and dead critters were buried in this mud. Ammonites, clams, and other shellfish from this layer are abundant fossils and have been collected since the 1860s. At Redbird, a fold in the strata alongside the eastern margin of the Powder River Basin is a structure known as the Old Woman Anticline. This up-arching fold managed to bring the entire thickness of the Pierre Shale to the surface. Having the entire 5,000-foot-thick sequence on one slope allowed U.S. Geological Survey paleontologist Bill Cobban to inspect it, all in one place.

What he found was an amazing sequence of stacked ammonite zones—layers ranging from a few to many feet thick, each containing its own discrete group of ammonite species. Cobban's lifetime of work has shown that this zonation stretches from Alberta to New Mexico. Everywhere the Pierre Shale is exposed, the same sequence of ammonite zones can be found.

The amazing thing about Redbird was how many of the zones were present in a single place. There were 19 of them exposed in 3,600 feet of shale, a veritable evolutionary logbook of the Cretaceous. The lower part of this section is also known for its abundant fossil fish and marine reptiles. Ken Carpenter from the Denver Museum of Nature & Science collected a lovely *Pteranodon* skeleton at Redbird when he was a college student. This flying reptile had a seven-foot wingspan.

As we drove north, the rocks were dipping into the basin and the Pierre Shale was to our right. To our left was a long sandy ridge, the Fox Hills Sandstone, deposits laid down on beaches and estuaries as the Pierre Sea retreated. Since the beds are tilted to the west, the Fox Hills makes a ridge that runs for miles. I kept trying to explain to Ray that we were driving along the same beach that we had seen back in Denver. Because the Rocky Mountains formed after the sea departed, each of the basins is ringed with a series of tilted beaches that were brought back to the surface by the uplifting mountains. He was trying to be a good student, but his stomach was already making him regret that we'd forgone the steak for a long, dusty drive.

We turned off to the west and onto ranch roads that seemed to go on forever. We lost track of the turns and gates. Finally, we drove through the Fox Hills Formation and into the Lance Formation, with its characteristic rusty sand beds. Here and there, the sand formed huge, elaborately swirled concretions that were as large as our truck. Some of these weathered out to form garish hoodoos in stark contrast to the conservative landscape. By now, the sun was low in the sky and the shadows were lengthening.

Eventually the road wound past some conical hills and we pulled down into a gully and into the area that Lee had been quarrying for years. Lee and his dog, a lazy golden retriever, got out, and it was immediately apparent that he was really at home there.

There were several levels where overburden had been scraped away, and fresh tan sandstone lay flat at ground level. Big brown bones were embedded in the sandstone everywhere we looked. We wandered around, and Lee was quick to tell us that almost every bone here was from an *Edmontosaurus*. We talked for a while about how you kill and bury an entire herd of duckbilled dinosaurs. The bones were isolated from each other, so whatever process it was had done a good job of spreading the skeletons around. Lee counts femurs as a way of counting individual carcasses because everybody, human or dinosaur, only has one left leg. So far he had counted 29 left femurs.

I was able to find a few fossil leaves, common species that Erling had described 60 years earlier. As Lee waxed on, Ray realized that we weren't going anywhere fast. He calculated the length of the drive back and realized that those steaks were never going to happen. Lee slipped him some beef jerky and kept talking. I rooted around in Big Blue and came up with a jar of peanuts. Peanuts and jerky don't equal steak, and Ray started to fade rapidly.

As the sun finally set, Ray's stomach prevailed over his interest in dinosaurs, and we said our good-byes to Lonesome Lee and the dead Eds.

We wound our way back out to pavement and headed north to Mule Creek Junction. There's not much

DUCKBILL
AND
DUCK

here at all. The Wyoming Highway Department maintains one of their environment-friendly rest areas at the junction, and one time I made the mistake of camping here. The water-free toilet system earned this place the name "Poopy Town," and I had one of my most unpleasant nights ever here, gaining relief from the smell and mosquitoes only when the wind blasted through.

Ray and I arrived in the town of Edgemont long after dark and well past the closing time of the town's dining establishments. In the end, our only option was the truck stop at the edge of town. It was one of those horrible places where a great diversity of fried products is displayed in a glass case, but it's not at all clear how long they've been there or what they are. We bought some cylindrical things that initially looked like fried spring rolls but ended up having some weird fossilized filling. Giving up on the local fare, we microwaved some frozen burritos, and I promised that I would do a better job of feeding the Troll. We finally found a crappy little room in the Rainbow Hotel and fought heartburn to fall asleep.

6
DINOSAUR TRACKS AT 65 MPH

The day that dawned was a huge improvement over the previous night. It was one of those High Plains sparklers with a dark blue sky, not a cloud above, and no hint of too much heat. The road from Edgemont climbs up a long pine-topped ridge and into the Black Hills of South Dakota. The Hills are magical, thick stands of lodgepole pine giving way here and there to open parklands of tall grass. Formed in the classic Rocky Mountain style when a giant area of buried granite was pushed up through a two-mile-thick slab of Paleozoic and Mesozoic–layered rocks, the uplift makes one of the more spectacular geologic bull's-eyes in the Rocky Mountain region. With giant concentric rings of geologic formations, each ring of rock younger than the one inside it, the Black Hills Uplift is what geologists call a breached dome. It was just another place where one of Dave Love's pigs had woken up under a blanket of formations.

The granite core of the Black Hills has been irresist-ible to the makers of massive monuments, with Mount Rushmore as the first huge effort and the endlessly ongoing Crazy Horse as a second helping. Both monu-ments were the obsessions of individual white men who ended up donating their lives to the projects. Every time I see these giant granite carvings, I can't help but think of the Sioux people who successfully bartered with the U.S. government in 1868 so they could keep the Black Hills while allowing settlers to pass around them peacefully. George Custer's arrogant Black Hills expedition of 1874 was the first step toward the demolition of this agreement, and the fact that his prospectors found gold sealed the deal. I don't know what it feels like to be Sioux, but I can only imagine that these giant rock carvings cause a lot of Lakota stomach acid.

We'd left Edgemont without so much as even a cup of coffee, the memory of the bad meal driving us quickly out of town. Highway 18 heads north up the ramping flank of the Black Hills toward Minnekahta Junction. This is the site of one of paleobotany's greatest failures. In 1892, a guy named F. H. Cole from the town of Hot Springs discovered an amazing deposit of fossil cycadeoids just south of Minnekahta. Cycadeoids, better known as bennettites, are often incorrectly called cycads, because they have similar trunks and fronds, and are some of the most charismatic of all fossil plants. Petrified trunks the size of fireplugs, they're often used in paintings of dinosaurian landscapes. In fact, it's rare to find a dinosaur painting that doesn't have one these overgrown pineapple–shaped plants somewhere. Their surface is patterned with diamond-shaped pits that are the remains of leaf bases, and embedded in the leaf bases are the remains of reproductive structures that are close to being true flowers.

The Minnekahta cycadeoids were petrified and agatized, which allowed them to be sliced open to expose perfect anatomy. One of Marsh's field men, George Wieland, got so excited by this anatomy that he devoted his life to studying it and eventually pub-lished the giant two-volume *American Fossil Cycads,* the first volume in 1906, the second in 1916 (at that time, cycadeoids were still called cycads). The Minnekahta site was so rich that Wieland made plans for a fossil plant visitor center and, using his Yale connections, was able to influence the National Park Service to protect the site. In 1922, President Warren G. Harding dedicated a 320-acre patch of land just south of Minnekahta as Cycad National Monument.

HMMM...

SAVVY BACKHOE DRIVERS SAVE THE DAY

Driving into town, Ray and I didn't see any signs for the national monument, the reason being that it's the only national monument to have ever been decommissioned. It's not clear whether it was rampant collecting (and Wieland did move hundreds of the fossil trunks back to New Haven) or simply a lack of anything to see, but in 1957, four years after Wieland's death, Cycad lost its national monument status. So much for the glory of paleobotany.

Digesting this sad tale did little for our growing hunger, so our first stop in Hot Springs was a local café that sported some very serious biscuits and gravy. We more than made up for the lost steaks of Lusk, slurped several cups of coffee, and talked about the Hot Springs mammoth site.

Located at the edge of town and looking for all the world like a giant gymnasium, the mammoth site is worth a long drive to see. In 1974, an alert construction worker named Porky Hanson was moving earth when his backhoe hit a chunk of what turned out to be mammoth bone. There are two types of heavy-equipment operators in this world: those who care about what they are digging through and those who don't. Fortunately for the Hot Springs tourist industry, Porky was of the former persuasion. Now Hot Springs is more famous for its mammoths than it is for its hot springs.

The discovery site was on the top of a hill at the edge of town. The rock beneath Hot Springs is the soft red Triassic Spearfish Shale, but that rock is underlain by a layer of limestone that dissolves away to make underground caves. Around 26,000 years ago, an underground depression collapsed, forming a 60-foot-deep hole in the slimy Spearfish Shale. The hole filled with water to become a lake, but it was a dangerous lake due to its slick and vertical shoreline. Then the fun began. One after another, dozens of mammoths let their curiosity override their caution and found themselves treading water. I don't know how long a mammoth can tread water, but I can tell you that it wasn't long enough, and one by one they drowned. Eventually, the lake silted up and filled the depression. And some 25,000 years later, a town was founded on the spot.

After the backhoe driver found the bones, word of the discovery made it to mammothologist Larry Agenbroad, who went to work with the town fathers to preserve and study the site for the next 40 years. Now it's a bustling building full of tourists and eager interpreters. More than 100,000 people visit the site each year. When Ray and I visited, more than 2,000 bones had been retrieved and the body count was up to 49 Columbian mammoths and three woolly mammoths, not to mention a smattering of wolves, coyotes, camels, and 22 other species of mammals that couldn't tread water forever.

We were amused to learn that most of the mammoth skeletons were young males. Because mammoths are close relatives of African elephants, speculative insights about mam-

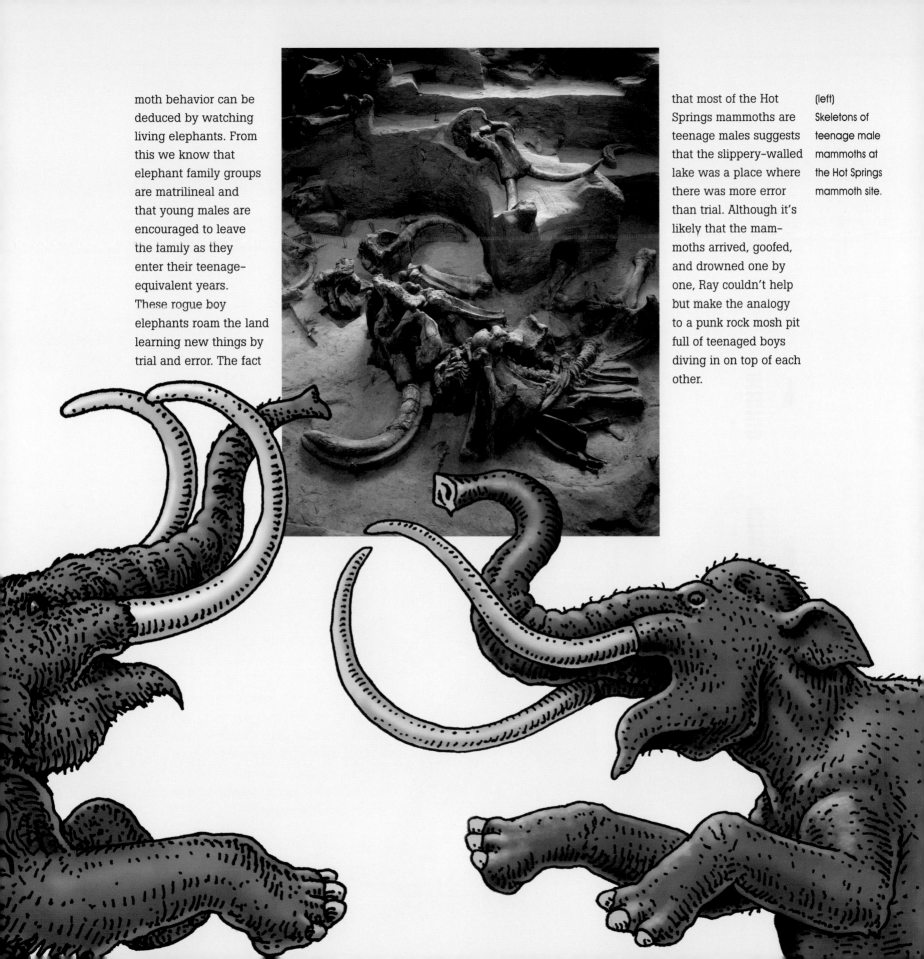

moth behavior can be deduced by watching living elephants. From this we know that elephant family groups are matrilineal and that young males are encouraged to leave the family as they enter their teenage-equivalent years. These rogue boy elephants roam the land learning new things by trial and error. The fact that most of the Hot Springs mammoths are teenage males suggests that the slippery-walled lake was a place where there was more error than trial. Although it's likely that the mammoths arrived, goofed, and drowned one by one, Ray couldn't help but make the analogy to a punk rock mosh pit full of teenaged boys diving in on top of each other.

(left) Skeletons of teenage male mammoths at the Hot Springs mammoth site.

In 1983, the mammoth site research team found the skull of *Arctodus*, the short-faced bear. Ray, who was already happy because of the biscuits and gravy and the mammoth mosh pit, could barely contain himself when he heard about it. Ray has a thing for massive predators, and *Arctodus* was one wicked bear. Standing nearly six feet at the shoulders when on all fours, this 1,400-pound bruin was a long-legged killing machine that would make a grizzly look like a sissy. Given the size of their limbs and the nature of the Late Pleistocene landscape, short-faced bears were probably open-ground predators that chased, caught, and ate bison and horses. A bear that can run down a horse on an open field is a scary concept.

From Hot Springs, it was downhill to the badlands of South Dakota. As we left town, we started to climb out of the Spearfish Shale and into younger layers of rock. "Climbing downhill" makes sense because we were heading across the east side of the Black Hills, where the sedimentary layers are also dipping to the east. We

*Arctodus,
the short-faced
bear, running
down a
paleontologist.*

passed by a tiny exposure of Jurassic rock and then into interbedded layers of sandstone and mudstone that, given their position above the Jurassic rocks, must have been Cretaceous. I had recently perfected a technique for spotting dinosaur tracks at 65 miles per hour. I told Ray about this and said that I would teach him the secret. He was skeptical.

Just then, we crossed a small bridge over the Fall River and were passing a road cut on our left when I saw the kind of rock that makes me stop and look for tracks. I edged Blue onto the grassy shoulder and hopped out. Ray, thoroughly unconvinced and barely interested, was changing the film in his camera, so I had a two-minute head start.

It didn't take two minutes. The third rock I saw had a giant toe imprint, and the next one had a spectacular seven-inch-long track. The problem was that the track had four toes, all pointing in the same direction and each tipped with a wicked claw. It was a track, but it wasn't from a dinosaur. By that time, Ray had made it across the road and was asking me if I'd found anything yet. I pointed at the splendid track and undid Ray. It was a real beauty, clearly defined on a 200-pound chunk of purple sandstone. He sat down next to it, dumbfounded, and stroked the impression of the large foot. We gazed at the trace fossil, trying to conjure up the creature that made it. "Maybe it's a footprint from a completely unknown dinosaur," Ray wondered out loud, the excitement rising in his voice. Later, it would become obvious that we were looking at the hind footprint of a big crocodile, but at the time we were still trying to make a dinosaur out of it. The surrounding rocks contained poor imprints of leaves, and I surmised that we were in the 100-million-year-old Fall River Formation. We later learned that our track was some of the first direct evidence of vertebrate life from this formation.

SPOTTING DINO FOOTPRINTS AT 65 M.P.H.

How to Find Roadside Dinosaur Tracks

People usually don't believe me when I tell them that I can spot dinosaur footprints from the freeway as I'm speeding along at 65 miles per hour. Then they're surprised when we pull over and find a slab of stone covered with giant three-toed tracks. The truth is, I don't see the tracks from the car. And no, I didn't plant them there. What I do is read road cuts the way you read billboards.

Finding dinosaur tracks from a moving vehicle is really a simple thing that more people should do. I stumbled on the technique back in 1983 when I found a bunch of bird footprints in a Paleocene outcrop in Wyoming. When you say fossil footprints, most people expect to see a flat rock surface with obvious dinosaur foot–shaped imprints. And sometimes you do see just that. Dinosaur Ridge west of Denver has a famous exposure of footprints right along the road, and thousands of people see them every year. But that's the rare condition. Far more common is for the track-bearing layer to be exposed in cross section. Think about it this way: a mouse walks across your birthday cake and leaves a nice little set of tracks in the frosting. Those tracks are pretty obvious when you look at the top of the cake, and if you saw them, you'd probably think twice before eating a slice. But if you had a piece of cake and only looked at it from the side, all

The giant Fall River crocodile track.

you would see would be subtle impressions of the toes, and you would probably eat the cake because you wouldn't recognize the impressions as tracks.

In the case of fossilized footprints, mudstone and sandstone take the place of the cake. Layers of these rocks were once layers of mud and sand. The mud was deposited as particles settled in still water, while the sand was deposited by moving water. Mud is funny stuff because it's squishy and slippery when wet but pretty hard when dry. Imagine a mudflat. Then have a few dinosaurs (or crocodiles) walk across it. After that, bake the mudflat in the sun until the mud dries and hardens. Then flood the mudflat with moving water and bury it in a foot-thick layer of sand. The footprint depressions will fill up with sand. When the sand turns to sandstone and erosion exposes the layers in cross-section, you'll be left with positive casts of the footprints projecting off the bottom of the sandstone layer.

Here's how it works: Sandstone is usually harder than mudstone, so when stream erosion cuts a valley and exposes horizontally layered rocks, the sandstone layers will stick out relative to the softer mudstone layers, which erode faster. Erosion exposes the sides of sandstone layers, and undercutting exposes the underside of the layers. So tracks and trackways appear as bumps projecting off the bottom of the sandstone layers. But there's a catch. As undercutting continues, eventually the slab of sandstone will fall off the cliff face, but it will almost always fall

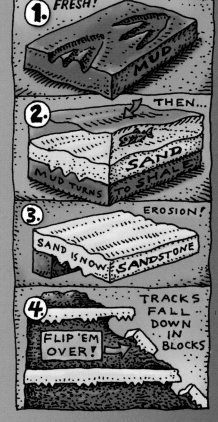

1. FRESH! MUD

2. THEN... SAND MUD TURNS TO SHALE

3. EROSION! SAND IS NOW SANDSTONE

4. FLIP 'EM OVER! TRACKS FALL DOWN IN BLOCKS

bottom side down. For this reason, most people don't find fossil footprints because they don't think to flip the slabs over. To find dinosaur tracks, all you need to do is look for thin sandstone layers (in Mesozoic rocks) that have odd projections on their undersides. This you can see from a moving vehicle. Stop the car and see if the projections are shaped like dinosaur feet. And remember to flip over fallen slabs. It's surprising how often this method works.

Not having a permit to collect fossils on South Dakota road cuts, we reported the find to the South Dakota School of Mines (SDSM), which came down the next week and hauled it back to their museum. We continued east, popping out of the Black Hills and onto the plains at an uninspiring truck stop just north of the Cheyenne River. The rocks that poked through the rolling, grassy hills were the classic dark gray shale of the Cretaceous Seaway. We were back in ammonite country.

We turned north and headed for Rapid City. It was near here in 1895 that George Wieland made a phenomenal discovery. He found the skeleton of a marine turtle that was as long as a Volkswagen Bug. This animal, known as *Archelon*, was nearly perfect except for its missing right rear flipper. Wieland surmised that the flipper had been bitten off by a hungry shark. He removed the turtle to Yale College, where it can be seen today in the Great Hall of the Peabody Museum. You can also see this turtle on the

Peter Larson in the jaws of *Acrocanthosaurus*.

side of certain U-Haul trucks. Someone at U-Haul must be an isolated paleonerd, because *Archelon* is not the only paleothemed U-Haul truck design. Others are decorated with Illinois's state fossil, the Tully Monster, while still others show the asteroid that killed the dinosaurs. On my next move, I'm definitely using U-Haul.

There was a second nearly complete *Archelon* found in the late '70s. It was discovered on the Pine Ridge Indian Reservation by a guy named Frank Watson, prepared by a guy named Peter Larson, passed on to a guy named Kirby Siber, and sold to the Natural History Museum in Vienna. I remember encountering this huge fossil in this classic European museum and wondering how such a fine resident of South Dakota had made its way to Austria. If you don't want to go to Austria to see it, there's a cast of this monster at Reptile Gardens near Mount Rushmore.

As we headed north, the sky grew increasingly dark and the air tasted smoky. A forest fire was burning in the Hills. Ray got nervous, and the dread in his voice grew as the radio started to warn about the possibility of evacuations.

Rolling into Rapid City through the relentless strip mall hell didn't do much to alleviate his feelings. To ease his angst, I called and found that a friend at SDSM was kicking around her office and would be happy to show us the collections.

South Dakota is amply endowed with fossils. Some of the world's finest Late Cretaceous dinosaurs, marine reptiles, and ammonites are here, as well as a splendid array of Eocene to Miocene mammals. The museum at SDSM is an old-style presentation with lots of fine skeletons collected over the years. I was surprised to learn that their *Edmontosaurus* had been collected by the Denver Museum in 1922 and had been involved in an elaborate three-way trade with Princeton University. Princeton apparently got an Eocene *Phenacodus* mammal skeleton, and I'm still trying to figure out what Denver got. Ray was in seventh heaven because of all of the marine creatures on display: a long-necked plesiosaur named *Alzadasaurus* from Iona, a huge *Xiphactinus* bulldog fish from six miles east of Hot Springs, a mosasaur from

Frederick, and a splendid *Squalicorax* shark from Fall River County.

Most of the staff were off working in Badlands National Park on a site they called the "Big Pig Dig." We got excited about this concept: a whole dig full of the angry entelodonts that we had been following across the countryside. They were finding two kinds of big pigs: *Megachoerus* and *Scaptohyus*. My friend Julia Sankey was holding down the fort while they were rooting away, and she showed us around the back rooms. The fossil lab was in a modified old gymnasium, and the collections were stored around the edge of an empty swimming pool. We spent the afternoon pulling open drawers full of bones and teeth and trying not to fall into the deep end.

Our day wasn't over yet, and we decided to drive up to Hill City to visit the Black Hills Institute of Geological Research. We got to Hill City around 10 P.M. that night and checked into a room. We rousted Pete Larson of *Archelon* fame out of bed for a beer at the Mangy Moose and chatted about the old times and the really old times.

Peter Larson is one of the most controversial figures in modern paleontology. He, his brother Neal, and his friend Bob Farrar run the largest and most successful commercial fossil operation in the country. Collectively, and with a tip of the hat to the Old West, Pete, Neal, and Bob are called the "Black Hills boys" or often simply "the boys." I've known Pete since 1982 and have shared a love of the Hell Creek Formation and the last days of the dinosaurs with him. Pete was born on the Rosebud Reservation and grew up collecting Fox Hills ammonites and the bones of sheeplike oreodonts from the White River badlands. When he was eight, he opened a fossil museum in a shed in his backyard. In eighth grade, he won the state science fair with a fossil exhibit. He went to college at SDSM and then started a fossil and mineral business with his buddy Jim Honert. Pete's big dream is to build a huge fossil museum in the Black Hills.

The Black Hills Institute grew on the backs of ammonites, but things got interesting when Pete got serious about dinosaurs. Pete found out about a bone bed near Faith, South Dakota, and soon leased the rights to dig the site, naming the quarry after the old ranch woman that owned the property. The Ruth Mason Quarry is a two-foot-thick layer of mudstone that's chockablock full of hadrosaur, or duckbilled dinosaur, bones. Pete mined them, collecting thousands. From his inventory, he began to assemble skeletons and sold them to museums around the world. Initially, the price for a delivered and assembled *Edmontosaurus* was around 300 grand. Not bad, when you consider all the work that went into excavating and assembling it and the fact that you were getting a real dinosaur.

On August 12, 1990, Pete and his crew were digging duckbills at the Ruth Mason Quarry. Pete's girlfriend at the time, Sue Hendrickson, decided to take a break from the tedious work of digging duckbilled dinosaurs with an X-Acto knife and took a stroll across the prairie to a distant clump of badlands. When she got to the muddy bluffs, she was confronted with the comfortable sight of a string of dinosaur vertebrae sticking out of the hill. There was something unusual about these bones, and she ran back to get Pete.

The dinosaur was a *Tyrannosaurus rex*, and it was a very nice one. Pete wrote a check for $5,000 to Maurice Williams, the landowner, and the boys dug like hell to remove the nearly complete skeleton from the ground in less than three weeks.

Once they got the skeleton back to Hill City and started chipping rock away from bone, they realized that they had a world-class find. Their lifelong plan to build their own natural history museum suddenly began to have some real potential.

But events didn't unfold as Pete had planned. A disgruntled acquaintance of Pete's contacted Williams and suggested that he had been ripped

Wes Wehr and the skull of *T. rex* Sue.

off. Williams was part Sioux and his ranch was within the bounds of the reservation, so the tribe began to take notice. Meanwhile, Vince Santucci, a paleontologist with a law enforcement degree who was working for Badlands National Park, also got interested in the reservation rex. Vince was worried about the illegal collection of fossils from national park and Native land and was suspicious about the activities of the Black Hills boys. It didn't help that a giant turtle from the reservation had ended up in Austria. With the landowner, the tribe, and the feds stirred up, a lot of different agendas started to crowd in on the simple concept of a natural history museum in Hill City, and a swarm of lawyers started getting paid to think about the legalities of dinosaur ownership.

On May 8, 1992, I visited the Black Hills Institute with my friend Wes Wehr to see what they were up to. They had just finished preparing the right side of the skull of the *T. rex* that they had named Sue and were preparing to ship it to Georgia for a CAT-scan. I photographed Wes next to the awesome five-foot-long head, and we headed back to Denver. Five days later, operating with a warrant from the state attorney of South Dakota, the FBI arrived in Hill City and seized Sue. They also seized other fossils and the business records of the Black Hills Institute. The city rallied in protest, but to no avail. The skeleton was plastered, crated, and hauled down the hill to a storage room on SDSM campus.

More lawyers got involved. Eventually, the government ruled that the original sale had not been valid and that the dinosaur should be auctioned to the highest bidder, with the proceeds going to Maurice Williams. Based on the records seized at the institute, a 39-count indictment was served on Pete, Neal, and Bob. Most of the counts were dropped, but the jury convicted Pete of failing to declare travelers' checks in excess of $10,000 when returning from Japan and sentenced him to two years in the minimum-security prison in Florence, Colorado.

Pete served 18 months and was released to a halfway house in Rapid City to serve the remaining six. I visited him in Florence when he was in month 12, and we spent a melancholy afternoon talking about *Triceratops*.

It was a sad situation. Even more so when the excavation of Sue wasn't even part of the conviction: her excavation itself had not been illegal. The feds had ruled that Williams, as a Native with some of his land held in federal trust to avoid property tax, was not allowed to sell any of his property, land, or dinosaurs without prior approval. Williams claimed that the $5,000 check from Pete labeled "for theropod Sue" was not a sale anyway, just a fee to prospect.

Meanwhile, Sue was scheduled for auction in New York. I visited Sotheby's a few weeks before the big day and wandered through a warehouse full of antique furniture to inspect the plastered skeleton. On October 4, 1997, Sue went on auction, and 90 seconds later she was sold to the Field Museum in Chicago for $7.34 million ($8.36 million with the commission). The Field had arranged financial backing from Disney, McDonald's, and a several supporters. The dinosaur went to Chicago and was unveiled in Stanley Hall at the Field Museum on May 17, 2000, before mesmerized crowds. Pete, living in a halfway house in Rapid City, missed the opening.

The next morning, after a hearty breakfast, Ray and I wandered over to the institute for a tour. We were greeted by Pete's younger brother, Neal, who is a full-on, raging ammonite fanatic. Words cannot describe how much he loves the *Scaphites, Baculites,* and other oddly twisted heteromorph

Neal Larson being attacked by a *Placenticeras*.

AMMONITES, TRILOBITES, AND ARROWHEADS

ammonites. If you start talking to Neal about anything, you'll soon be talking with Neal about ammonites. Ray and I like ammonites; I'd even say we love ammonites. They're part of the holy trinity—ammonites, trilobites, and arrowheads—that shaped my adolescent fantasies of discovery. But if you want an overdose of ammonites, go to Neal.

In the last couple of years, Neal had been sitting at the feet of Bill Cobban and begun to learn the real science of ammonites. Cobban, at the time a 90-year-old USGS paleontologist in Denver, was the reigning Zen master of ammonites. A thin slip of a man, Cobban had a flawlessly photographic memory. He could remember, with precision, sites that he collected at 60 years ago. Since the Pierre Shale is widely exposed over the Great Plains from New Mexico to Canada, Bill had collected hundreds of thousands of ammonites from nearly 14,000 localities. Neal would like to collect millions of ammonites, and he's well on his way. Neal regaled us with tales of *Solenoceras*, *Placenticeras*, *Spiroxybeloceras*, and *Sphenodiscus*; he

attacked Ray with a giant plastic *Placenticeras*; he told us about ammonite jaws and how holes made by limpets in ammonite shells can sometimes be mistaken for mosasaur bite marks. After a couple of dizzying hours, he passed us off to Bob Farrar, who took us into the storage room and showed us parts of dozens of dinosaurs. By this time, Ray was in a rhapsodic and blissful state.

Later, interviewing Pete, I tried to learn how many dinosaur skeletons he had collected. Not counting the inventory of enough bones to make 50 or so *Edmontosaurus* skeletons from the Ruth Mason Quarry, he ran off a list of at least a couple of dozen other skeletons and skulls, including a several *Triceratops*, an *Ornithomimus*, and a huge *Acrocanthosaurus*. And in the years since Sue came and went from his life, he's collected half a dozen more *T. rex*. Regardless of how history treats him, Pete Larson will go down in the books as one of the most prolific dinosaur hunters of all time. And he's still dreaming of his museum in Hill City.

7
OUT IN THE GUMBO

Gumbo is a northern plains name for a reviled badlands bedrock composed of equal parts mud and volcanic ash. My favorite road to the gumbo is South Dakota State Highway 85, which heads straight north from Belle Fourche. It's an amazingly straight road that parallels the course of the Little Missouri River and runs right past the geographical center of the United States. From Belle Fourche to Buffalo, you drive on rolling hills carved into the Pierre Shale, that giant pile of mud left by the second-to-last sea to cover North America. The Pierre Sea slowly slid off the continent 68 million years ago, and new land formed in its place. Heading north, you cross the Fox Hills Sandstone, remnants of the ancient shoreline, and when you reach Crow Buttes, you've arrived at the Hell Creek Formation. I love the Hell Creek Formation; the first distant sight of these lonely buttes always feels like a homecoming to me.

Crow Buttes stick out of the prairie like two huge piles of melting ice cream. The historical plaque next to the brand-new roadside mini-mart tells the gruesome tale of a raiding party of Crow who were trapped on these buttes by a larger party of Sioux. The Crow were surrounded, without water, and eventually were killed before they had the chance to die of thirst. Weathering out from the buttes are bones of dinosaurs that died on the same spot 66 million years before.

The Hell Creek Formation is the source of some of the best-known dinosaurs in the world: *Tyrannosaurus rex*, *Triceratops*, and the flat-headed duckbill *Edmontosaurus*. As flat as a pancake, the Hell Creek Formation is of decent size. It's about 300 feet thick and stretches from the central Dakotas to central Montana and from southern Canada to the Black Hills. At the Montana-Wyoming border, due to the odd provinciality of geology, the Hell Creek Formation changes its name to the Lance Formation, which thickens as it extends south to Casper.

North of Crow Buttes is the tiny town of Redig, which has one family and more than 300 junked cars. A bit beyond that, the broad valley of the Grand River opens up and there is a sign that, illogically, points north to Buffalo and east to Bison. On the eastern horizon are the Slim Buttes, a strip of tree-covered high country formed by a resistant cap of Eocene and Oligocene rock lying on top of the softer Hell Creek and Fort Union formations.

Near Slim Buttes on September 9, 1876, just more than two months after the Battle of the Little Bighorn, Captain George Crook's men surprised a band of Sioux under the command of a Miniconjou chief named American Horse and began the process of exacting revenge for the slaughter of Custer's 7th Cavalry. American Horse and his people had been camped at the base of the Slim Buttes and were surprised by the soldiers in a dawn raid that laid waste to the encampment and decimated and scattered the Sioux. After the battle, Crook and his men, already

exhausted by two months on the trail, spent a brutal week slogging from Slim Buttes to the town of Lead in the Black Hills. It rained the whole way, and the water soaked the Hell Creek Formation and Pierre Shale into a slimy, muddy mess of gumbo hell. The soldiers were low on food, and the "horse-meat march," as it came to be known, was made miserable by the antisocial behavior of the 70-million-year old rocks.

A bad rain does the same thing today, and woe is the Rocky Mountain roadster that ventures onto the gumbo on a wet day. Volcanoes erupting near Livingston, Montana, in the Late Cretaceous dropped airborne ash into the Pierre Seaway and later onto the Hell Creek landscape. This ash has weathered into a clay called bentonite, which is as slick as axle grease when wet. This clay contributes to the look of the modern landscape. The Hell Creek forms buttes of many sizes, including little haystack buttes that lie like loaves of bread on the landscape. The Pierre Shale tends to weather flat, exposing itself only in steep-walled gullies or cuts along the roadside.

The town of Buffalo in Harding County, South Dakota's northwestern corner, bills itself as the *T. rex* capital of the world, and it's not an idle claim. Of the forty or so *T. rex* skeletons ever found, Harding County can claim at least seven. Most have been collected on private ranches by commercial dinosaur diggers. *T. rex* skeletons are such a big deal that they usually get their own names. One Harding County *rex*, known as Z-rex, spent   some embarrassing time on Ebay but only received

Running with *Anzu*.

bids from overheated dino kids who couldn't back up their multimillion-dollar bid. Stan, a big *rex* excavated by the Larson brothers in 1992, is on display in Hill City along with another *rex* known as Duffy, which the Larsons found less than a mile from Stan. Stan was named for Stan Sacrison, a Buffalo native and eagle-eyed bone picker. Stan's brother, Steve, also found a *rex*, named, you got it, Steve. And so it goes, 66-million-year-old carcasses named after the guys who found them.

The Black Hills boys had hoped that *T. rex* Stan would fill the hole left by *T. rex* Sue. Looking back from 2022, things did not pan out as planned. Pete and Bob split up with Neal in 2007. More lawyers got involved and *T. rex* Stan was auctioned in 2020 for the eye-watering sum of $31.8M. Stan is now in Abu Dhabi. Pete and Neal still live in the Black Hills.

Mike Triebold is another commercial bone digger who hangs his summer hat in Harding County—and his winter hat in Woodland Park, Colorado. Mike runs a smooth operation, prospecting for patches of badlands from his small plane and living in a comfortable trailer on ranches that he leases to find bones. He isn't so interested in big dinosaurs; he'd rather find the small and rare ones, dinosaurs that are worth a lot more per pound than a giant *rex*. In the 1990s, Mike had a lot of success in Harding County, selling a nearly perfect duckbill skull to a museum in Fukui, Japan, and a whole collection of dinosaur spare parts from one amazing bone bed to the National Science Museum in Tokyo. Working with another collector, Mike and his team excavated a pair of oviraptor-like skeletons. Fragments of this animal had been found in the Hell Creek before, but no one had ever bagged a whole one, much less a matched pair. Oviraptors are well known from Mongolia, where they have been found fossilized in place on nests of their own eggs. The name oviraptor, or "egg-stealer," is really a misnomer, since it was based on the mistaken impression that the animals had been fossilized while stealing eggs, not sitting on them.

Oviraptors are hideously ugly beasts with contorted birdlike faces and disproportionally long arms with huge hooked claws. The Mongolian oviraptors, like the Mongolian velociraptors, are small animals, waist-high to a normal guy. Triebold's Harding County oviraptor-like beast is more than twice that big. At seven feet with three-foot arms and four-inch claws, this is one creepy, unsettling animal. Both specimens were purchased by Carnegie Museum in Pittsburgh, where they're part of the newly renovated dinosaur hall. Once they were in the hands of a museum, they could be formally named, and now the Harding County oviraptor is known as *Anzu wyliei*.

In southern Harding County, another Mike, Mike Hammer, found a small dinosaur that had a heart of stone, literally. The animal, a small beast known as *Thescelosaurus*, is about the size of a big deer (with a crocodile-sized tail). Mike sold the skeleton to the North Carolina Museum of Natural Sciences in Raleigh, where paleontologist Dale Russell CAT-scanned it. There, in the upper part of the intact rib cage, was a lump of ironstone the size and shape of a heart. Russell and his large-animal veterinarian friends, who were used to looking at X-rays of horses and cows, came to the utterly amazing conclusion that they were looking at a petrified dinosaur heart. Debate rages over this interpretation, but the lump is the right size and in the right place. Studies of modern decaying carcasses show that fatty organs rot faster than those with less fat, and it might just be possible that this little dinosaur got buried fast enough that the blood in its heart catalyzed iron carbonate molecules in the sediment, which fossilized the heart.

Ray and I sailed through Buffalo, pausing only to grab a picture of the rotting sign that proclaims the town's claim to *T. rex* fame. Then we passed the Cave Hills, great wooded buttes capped by Paleocene sand layers that were deposited by the beaches of the last Western Interior Sea. About a million years after the last dinosaurs, North America's last saltwater soaking, known as the Cannonball Sea, covered the center of the Dakotas and stretched out to the Arctic Ocean. Full of sharks and fishes, it conspicuously lacked the ammonites and giant marine reptiles of the Pierre Sea. These creatures had checked out when the dinosaurs went extinct. The end of the Cretaceous, commonly called the K-Pg boundary, is a tale told best by the badlands of southwestern North Dakota, and that was where we were headed.

We pulled into Bowman, North Dakota, just as the long summer day was ending, and I knew enough to head straight to the local sports bar, where we found a stalwart band of volunteers from the Pioneer Trails Regional Museum. We were soon enjoying fried cauliflower and Coors Lights with Dean Pearson and his team that included ranchers, a nurse, a dentist's wife, and a couple of couples who had the sense to retire somewhere interesting. This unlikely crew has built a strong research program and a beautiful exhibit museum under the guidance of Dean and his formidable mom, Dorothy. Dean's day job is supervising the mixing of cow cakes at the feed mill in nearby Scranton. By night and weekend, he's one of the premier amateur paleontologists in the country. He grew up in Bowman, a small town, but he had a big brain and his parents piled on the books. Now Dean writes them. Not only does he do credible paleontology, but he has also supervised archaeological excavations of the trails and camps left by Custer's 1874 expedition to the Black Hills and the plains to the north.

Thescelosaurus, the little dinosaur with a heart of stone.

We were in a tiny bar in a remote corner of one of the most rural and maligned states, and Ray couldn't have been more surprised as the beer mixed with theories of dinosaur extinction and methodologies of cataloging fossils.

The next morning, we drove west on Highway 12 through rolling plains. About five miles west of Rhame, the prairie abruptly fell away, exposing an endless confusion of badlands. I stopped the truck at the lip and welcomed Ray to the little-known badlands of the Little Missouri River. This was holy ground for me. The Little Missouri River begins near Devils Tower in northeastern Wyoming and flows a meandering course due north until it meets the Missouri River in what is now Lake Sakakawea. It's one of the least-known rivers in the United States, and only three tiny, obscure towns—Camp Crook, Marmarth, and Medora—grace its banks. The Little Missouri badlands were formed during the ice ages as runoff from the Black Hills eroded the soft bedrock of the Pierre Shale, Fox Hills, Hell Creek, and Fort Union formations. Before the ice ages, the Little Missouri River flowed all the way to the Arctic Ocean. The massive ice sheets of the Pleistocene stopped that nonsense, and now the Little Mo flows north only as far as the Missouri River.

In 1883, it was to these rounded mounds of soft rock that Teddy Roosevelt, the last American rex, retreated after the nearly simultaneous deaths of his mother and first wife. As a 24-year-old New York patrician, Teddy acquired a ranch and became, for a while, a frontiersman. It was in these hills that he shot one of the last wild bison on the northern plains and, paradoxically, began to embrace the concept of conservation. After only three years in the badlands of the Little Missouri, he returned to the East and politics, but his experiences in the valley influenced the rest of his life.

One of the least known and most beautiful national parks is the one named

for Theodore Roosevelt. It consists of two pristine regions of Little Missouri River badlands close to where the Montana–North Dakota line is crossed by Interstate 94. The badlands of the park are composed of the Sentinel Butte Member of the Fort Union Formation. Sixty million years ago, these rocks were mud and sand in a swampy lowland that looked like Georgia's Okefenokee Swamp. Like the modern Georgian swamp, the ancient Dakota swamp hosted turtles, alligators, and frogs.

Bruce Erickson at the Science Museum of Minnesota in Saint Paul has spent the better part of his career digging into a single hill near Theodore Roosevelt National Park to make an excavation known as the Wannagan Creek Quarry. Based on Bruce's work, we now know that this place was once a crocodilian Camelot. In addition to many types of turtles, more than six species of alligators, crocodiles, and crocodile-like champsosaurs lived in the same spot—more types of crocodilians than inhabit any modern swamp or river. Conspicuously absent from this reptilian paradise were dinosaurs.

The dinosaurs became extinct at the end of the Cretaceous Period, 66 million years ago, about 5 million years before the Wannagan Creek Quarry was the bottom of a swamp. If you travel upstream (to the south) on the Little Missouri from Wannagan, you descend through the stratigraphic column. About where Teddy killed his bison, you come to the transition from the Fort Union to the Hell Creek Formation. This stratigraphic level was the land surface when the dinosaurs bit the dust. Here, Cretaceous rocks are overlain by Tertiary rocks, forming the famous K-Pg boundary. The *K* was substituted for *C* because another time period, the Carboniferous, had already claimed that letter. More recently, the Tertiary Period was retired and replaced by the Paleogene Period which includes three epochs: the Paleocene, Eocene, and the Oligocene. So, the C-T became the K-T which became the K-Pg.

FORT UNION

THEN THE GIANT ROCK FROM OUTER SPACE SMACKED INTO THE EARTH.

CRETACEOUS

Pg

K

HELL CREEK FORMATION 66 MYA.

RAY TROLL '06

Ground zero for the study of the K–Pg boundary is the shrinking riverside town of Marmarth, the southwest-ernmost town in the state. Founded in 1908, Marmarth was home to a major hub of the Milwaukee Railroad and the first motion picture theater west of the Mississippi. With a population of 150 ranchers, railroad retirees, and High Plains drifters, Marmarth is the largest town in the smallest county in the nation. Small in population, that is. It's a 50-mile drive to the sparsely populated county seat of Amidon, whose 24 souls include the county's one law enforcement officer. Marmarth would be a good place to rob a bank … if it had one. It's been home to my own research on fossil plants and the K–Pg boundary since I first visited late in the summer of 1981, when as a student, I learned how to be a geologist. Marmarth, named for sisters Margaret and Martha, has a very special something that has drawn me back summer after summer for more than 40 years.

The buttes of the Little Missouri River expose the flat-lying layers of the last Cretaceous landscapes, and these hills are loaded with the bones of the last dinosaurs. Dean Pearson and his team have found more that 10,000 fossils representing more than 60 species of dinosaurs, mammals, birds, crocodiles, turtles, fish, amphibians, lizards, and pterosaurs. These finds include a partial *Tyrannosaurus rex* that Dean, in his opposition to the commercialization of paleontology, refuses to name (although a dinosaur named Dean has a certain ring to it) and parts and pieces of several *Triceratops* and *Edmontosaurus*.

Over the last 40 years, I've brought teams of urban volunteers to Marmarth to work with Dean's homegrown crew. Together we've dug more than 150 quarries in the Hell Creek and Fort Union formations and have collected more than 25,000 fossil leaves. Both Dean and I have obsessive-compulsive disorder when it comes to collecting large numbers of fossils, but the result is a pretty clear view of an ancient world.

When Ray and I rolled into town, summer was ending and Marmarth was as deserted as ever. The local historical society maintains the old Milwaukee Railroad bunkhouse, and 12 bucks buys you a clean, Spartan room with a guaranteed 4 A.M. train rumbling through. Evelyn Lecoe, the live-in host, welcomed us to town, and we dropped our gear and headed across the dirt-road main street to the Pastime Bar.

When I first came to Marmarth, the bar was owned by a guy named Shirley and run by a wanderer named Mike Luten. We called the place Rootin' Tootin' Luten's and knew that with the law 50 miles away, the bar could stay open all night.

Luten and the bar regulars called us bone pickers, but we called ourselves leaf diggers. Luten, a bit of a bone picker himself, showed us a *Triceratops* femur that he'd found sticking out of a gumbo butte. We showed him how to use plaster of paris and burlap strips to make a plaster jacket for the bone so he could carry it back to the bar.

Later that night, actually much later, the bartender from the lone bar in Rhame rolled in for a nightcap. He didn't need one, and the one he got put him out on the floor by the jukebox. Luten looked at me; I looked at my assistant, Rob; he looked at Luten's wife, Teresa; and she looked at the passed-out bartender. Then we all looked at the leftover plaster and burlap and, simultaneously, had a very bad idea. Twenty minutes later, a drunken bartender learned, for the first time, what it meant to be truly plastered.

Now the Pastime has a surprisingly great restaurant out back. It's a good thing, because my favorite spot, Mert's Café, had closed a few years before because of a leaky roof and a failed inspection. (In the early 1980s, Mert's had so many varieties of fried food products that my visiting Ph.D. adviser called it the "Loire Valley of the fried potato.") Troll and I murdered a couple of big T-bone steaks at the Pastime and headed to the buttes.

It seems that the settlers of the American West were

PREHISTORIC
LEAF LUST-
HARD TO
EXPLAIN

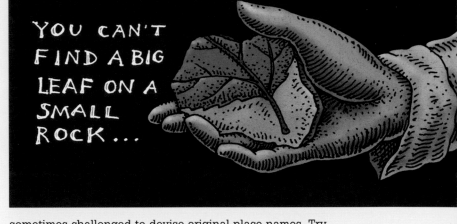

YOU CAN'T FIND A BIG LEAF ON A SMALL ROCK...

sometimes challenged to devise original place names. Try counting the number of Cedar Mountains or Cottonwood Creeks. The place we went to is known quite uncreatively as Mud Buttes. I first saw Mud Buttes in 1988 with Dean. It's a wonderland of eroding Cretaceous and Tertiary mud and sand, a thickness of about 50 feet of Cretaceous rocks overlain by 150 feet of Paleocene rocks. It's a great place to look for the last fossils of the Cretaceous and the first fossils of the Paleocene.

I wanted to give Ray the joy of finding dinosaurs, and Mud Buttes is loaded with chunks of ancient carcasses. The good news is that bones were everywhere, the great majority of which were scattered bits and pieces. Paleontologists have a name for tumbled chunks of unidentifiable dinosaur bones: *Chunkosaurus*. The fragmentary nature of the fossil record was a surprise for Ray, who was expecting perfect bones, not tumbled chunks. The bad news for Ray was that the land is overseen by the Bureau of Land Management, and it's only legal to collect bones with a permit. We didn't have one, so Ray's first dinosaur bone, a *Triceratops* vertebra, is still lying where he found it. For a guy who grew up dreaming of dinosaurs, this was a bitter pill to swallow. He sulked for a while and pleaded with me to bend the rules, but I was having none of it.

Mud Buttes is also one of the very best places on Earth to try to answer the question "What killed the dinosaurs?" I've spent a lot of time beating on this question, and my approach has been a simple one: find lots of fossils below the K-Pg boundary and lots above, then compare them and see how different they are. This

The End of the World as We Know It

In 1980, Walter Alvarez, his dad, Luis, and colleagues Helen Michel and Frank Asaro discovered that a rare platinum-group metal known as iridium was anomalously abundant in a finger-thick layer of clay at three K-Pg boundary sites in Italy, Denmark, and New Zealand. Using this tiny bit of evidence and a lot of chutzpah, they suggested that the iridium came from an asteroid that hit the Earth and killed the dinosaurs. The scientific world chortled and suggested that abduction by little green men was a similarly valid suggestion, but the Alvarez men were on to something. Within a year, dozens of new K-Pg sites were discovered, some in rocks that had been deposited under seawater and some that had been deposited in swamps and lakes. In all of these places, chemists talked to paleontologists who showed them the precise stratigraphic level where diverse Cretaceous microfossils (plankton fossils for marine rocks; plant pollen and spores for freshwater rocks) gave way to nondiverse Paleocene microfossils.

In 1988, my longtime science buddy, USGS fossil pollen specialist Doug Nichols, had sampled the rocks at Mud Buttes, and his results showed that Cretaceous pollen was found only in the Hell Creek Formation and not in the over-lying Fort Union Formation. This jibed with Dean's bones. He found abundant dinosaur fossils in the Hell Creek Formation but only turtles and crocodiles in the Fort Union Formation. Together, the plant and animal fossils point an accusing finger at an asteroid as the cause of the big extinction.

approach is predicated on the idea that I can recognize the K-Pg boundary when I see it, which turns out to be easier said than done.

Ray and I dug a foot-deep trench that crossed the contact of the Hell Creek and Fort Union formations. Our intent was to find the precise layer that marked the debris from the dinosaur-killing asteroid. Usually this layer is subtle to the point of invisibility, and you can't even tell you have it without doing costly laboratory analyses. But Ray's luck was good, and the trench we dug had a strange tan layer. As we got down on our hands and knees to inspect it, my eyes focused and I realized the layer was composed of tiny round balls of clay about a half a milli-meter in diameter. In a few rare K-Pg boundary sections in the Caribbean, the layer is composed of tiny glass beads known as microtektites that formed when molten droplets of target rock from the impact crater cooled as they flew through the air. On that day at Mud Buttes, Ray and I found microtektites that had been blasted all the way from the Chicxulub crater on the Yucatán Peninsula in Mexico to the southwestern corner of North Dakota. Sixty-five million years of weathering had reduced the glass to clay, but the characteristic teardrop and dumbbell shapes that are diagnostic of microtektites were clearly there.

After more than 20 years of work in North Dakota, this was one of my most exciting and most unexpected moments. We set the camera on self-timer and proudly posed by the tan layer of death. Later, lab work docu-mented the presence of anomalous amounts of iridium in this layer and confirmed its origin as asteroid debris.

There are still some scientists who doubt that the dinosaurs met their demise at the fiery fist of an asteroid impact, but I don't think many of the doubters have had the profound experience we did of walking in a valley full of dinosaur bone and then finding the distinctive K-Pg boundary layer and realizing that the layers above didn't contain a single fragment of *Chunkosaurus*.

Back in Marmarth later that night, we stopped by to visit a local high-school wunderkind, Tyler Lyson. Tyler's uncles own a lot of gumbo, and Tyler, a lean, laconic 17-year-old, had spent a few summers learning how to

The K-Pg Boundary Washing Machine

In 2008, eight years after Ray and I found the K-Pg boundary spherules at Mud Buttes, two commercial paleontologists, Steve Niklas and Rob Sula, working on a private ranch just two miles away, found a very unusual site. Fossil fish are quite common in the Hell Creek Formation, but they usually come as spare parts like bones, teeth, and vertebrae. It is extremely unusual to find a whole fish. Steve and Rob found a jumbled logjam of whole big fish. There were three kinds—bowfin, sturgeon, and paddlefish—and they were embedded in the sediments of an ancient stream bed that was about 30 feet below the K-Pg boundary. Steve and Rob were hunting for dinosaurs and had no use for the fish, so they collected a few specimens to give to museums. Then, in 2012, they passed the site off to a graduate student at the University of Kansas named Robert DePalma. I bumped into Steve in the field a few years later and he showed me the site. To me it looked like a volcanic ash fall had choked the stream and killed the fish. Boy was I wrong!

DePalma, who had secretly started working in the site in 2013 had a different idea. He surmised that the site was a very unusual expression of the K-Pg boundary. Rather than simply being a finger-thin layer of spherules, he proposed that this site represented a place where the seismic wave from the Chicxulub crater had created an earthquake that churned up a riverbed a few minutes before the incoming spherules started landing. He announced his findings at a scientific meeting in 2017 and I can remember sitting in the audience stunned by the audacity of his claims. In 2019, he published a paper that documented some of his claims and convinced me that the site was a very weird and cool example of the K-Pg boundary. Since fossils are formed by death and burial, this site is an example of a spot where an asteroid-impact induced earthquake created a local patch of quicksand that killed and buried a whole group of animals on the very day of the impact.

dig bones with a commercial digger. Quickly figuring out that finding dinosaurs isn't rocket science and that nobody needs a middleman, he started hunting bones on his own. His dad turned over the family workshop, and Tyler began to find amazing fossils. First it was a turtle jam: 35 turtle skeletons in an ancient streambed.

Then, not one but two *Thescelosaurus* dinosaur skeletons. Pretty soon he was hotdogging it: one night he showed us a hollow *Triceratops* horn full of baby *Tyrannosaurus* teeth. When I mentioned that I had always wanted to find a good *Triceratops* skull, he said that he'd seen dozens and took me for a hike where I found my own. By the time he was a senior BMOC at tiny Baker High School, Tyler had a regular stream of paleontologists making pilgrimages to his shop. A few years later, when he was a college student, Tyler made the find of a lifetime. He and his dad drove to Denver and showed me pieces of perfectly preserved dinosaur skin. Tyler had found a whole mummified *Edmontosaurus* hadrosaur. Dinosaur mummies are extremely rare, and they are extraordinarily hard to collect. With a typical dinosaur skeleton, all you need to do is chip away the rock to expose the bone and then carefully extract the bone, often in a plaster jacket. Individual jacketed bones don't weigh that much so it is possible to collect a dinosaur in a bunch of manageable pieces. Dinosaur mummies present a totally different problem because the skin is preserved as imprints in the rock, and you can't chip away the rock to expose the bone without destroying the skin. That means that you must collect the whole dinosaur on one piece, or at least in a couple of large pieces. And dinosaurs are big. Fortunately for Tyler, one of his brothers is a welder so the proper equipment was in the family. With this help, Tyler was able to collect the whole animal. The fossil was nearly 20 feet long and weighed more than 6 tons! Not bad for a college kid. Breaking with tradition, Tyler named his mummy "Dakota."

People who make big discoveries often find themselves with a lot of new friends. When the word of the mummy got out, a lot of people wanted to meet Tyler. And almost all of them had big plans for Tyler's dinosaur. Savvy beyond his years, Tyler parlayed the find into a book, a National Geographic television show, a deal for the dinosaur to become a North Dakota State Treasure, and a plan to endow paleontology research in Marmarth. Today, the mummy named Dakota is on display at the North Dakota Heritage Center and Museum in Bismarck and Tyler is the president of the well-endowed Marmarth Research Foundation. He's also a curator at the Denver Museum of Nature and Science.

Most pictures that portray the landscapes of *Tyrannosaurus*, *Triceratops*, and *Edmontosaurus* do a poor job of it, focusing on the animals but not the plants or the ecosystem. Often, the only plants in the scene are distant conifers behind a field of brown, pounded-down dirt. I call this misleading iconography "Monkey Puzzles and Parking Lots" because in these paintings, it commonly looks like dinosaurs are wandering around a parking lot fringed by Chilean conifers known as monkey puzzles. When I quizzed artists as to why they were painting bare earth, they told me that paleobotanists had forbidden them from using grass because it didn't evolve until the dinosaurs had gone extinct. Remove the grass and you're left with bare earth. This prompts the question "What was the ground cover in the Cretaceous?"

I took Ray to another spot near Marmarth that I discovered in 1986. Here we dug into tilting mudstone layers that used to be a mud bar on a big lazy Cretaceous river, and we found the remains of a meadow. This new site was full of information about the ground cover, and what we found was surprising. Instead of ferns, cycads, or horsetails, we found herbaceous flowering plants. Many we didn't recognize, but others we could, including buttercups, nettles, hops, and even something that looked suspiciously like marijuana. On closer inspection, the marijuana-like leaf had features that suggested it might also be related to hops.

Hops (genus *Humulus*) and marijuana (genus *Cannabis*) are the only genera in the botanical family Cannabaceae. This is an obscure fact often missed by connoisseurs of the products of either of these two widely imbibed plants. The fact that the last dinosaurs might have been browsing on an extinct missing link between hops and marijuana led to some fertile conversations about the real cause of their extinction. And it led to many more conversations about what to name these fossils.

One of the fun things about being a paleontologist is getting to name a new fossil species. If you can demon-

Monkey puzzles and parking lots.

A deceptive fossil herb from a Cretaceous meadow.

strate that nobody has previously named a specific fossil, then you can do it yourself by publishing a careful description and illustrations in a scientific journal, provided you pass the scrupulous review process. Troll was keen that this fossil get an appropriate name, one that reflected the widespread use and abuse of its descendants. But there was a problem. Since 1991, I've been enjoying the malted beverages of a Denver establishment known as the Wynkoop Brewing Company. One night several years ago over a free pitcher of Wynkoop's finest, I'd signed a contract with the bar's owner, on a napkin, of course, promising to name this Cretaceous hops ancestor after the brewpub.

Ignoring my obligation, Ray was adamant that the fossil be given a different name, one that reflected the plant's apparent affinities with marijuana. He was so wed to this idea that when he and I later attended a Ziggy Marley concert in Philadelphia, he set up a meeting with Ziggy so that he could make his case with a true ganja expert backing him up. Ray bribed a stagehand, and we found ourselves standing in a small room backstage with the man himself. Troll explained that I was a paleontolo-

gist and that we had great respect for Ziggy's dad, Bob Marley. Then he told Ziggy about the Cretaceous marijuana missing-link plant from North Dakota. Ziggy seemed puzzled by the concept of paleontologists backstage, but he was a surprisingly fossil-savvy guy. In retrospect, it was apparent that he knew that *Australopithecus afarensis*, the 3.5-million-year-old hominid fossil known as Lucy, was, like Haile Selassie, leader of the Rastafarians, from Ethiopia. He asked me, "Where de first mon from?" I was starstruck, didn't clearly understand his accent, and thought he had asked me where my fossil came from. I answered with two words: "North Dakota." The room was suddenly stone-cold silent. I looked at Ray, he looked at me, and I realized that I had just dissed Haile Selassie and Lucy. Fortunately, Ray used his backstage ways to smooth out the misunderstanding, and we left a room full of slightly confused Rastafarians with the concept that maybe, just maybe, we might name a fossil plant after Ziggy's dad. In the end, the marijuana-like fossil turned out to be a lobe-leafed nettle, a disappointment to both the bar owner, the Rastafarian, and Ray.

As always, I was sad to leave North Dakota. The drive from Marmath to the Montana line is a quick 15 minutes, and 10 minutes after that we rolled into the oil field town of Baker. The road south from Baker is the world's longest 34-mile drive. It consists of two 17-mile ruler-straight stretches. No matter how fast you drive, this road takes forever. To the west of the road is Medicine Rocks State Park, a maze of sandstone towers, spires, and pillars that are the remains of Paleocene stream channels. At the end of the road is the quirky town of Ekalaka, named for the Sioux wife of the town's founder.

Ekalaka is the home of the Carter County Historical Society and its surprisingly wonderful museum. The museum has a long history and some amazing displays. The first time I visited, I met the director, Marshall Lambert, a high-school teacher turned dinosaur digger. Marshall and his local supporters built a museum out of jagged chunks of petrified wood and filled it with skeletons from the Hell Creek Formation. This museum has the only *Anatotitan* (a giant duckbill) skeleton outside of New York City and one of the finest *Triceratops* skulls ever found.

The badlands around Ekalaka have hosted bone hunters for years. The Cleveland Museum of Natural History has a splendid little tyrannosaur skull from this area that was renamed *Nanotyrannus*, "the microscopic ruler," by Bakker. Rumor has it that the beautiful little skull was perched on a hoodoo when it was found by Cleveland curator David Dunkle in 1942. Hoodoos are pillars of sedimentary rock capped with something more resistant to erosion than the rock of the pillar. Typically, the cap is just a zone of hardened rock, but in fossil-rich areas, fossils can cap the hoodoos. Collecting the skull required nothing more than strolling up to the hoodoo and snapping the skull free. Paleopurists dream of being the first to prospect an area, because the fossils will litter the ground and cap the hoodoos.

Ekalaka also produced one of the first skulls of the bone-headed *Pachycephalosaurus*. There's a great old newspaper clipping in the Carter County Museum that shows a

cartoon of *Pachycephalosaurus* with a coconut bouncing off his bony head. It's still a mystery why these animals had skulls that were capped with bowling ball—thick wads of bone, and I've always liked the absurd idea that the skulls were for protection from falling coconuts.

Over the last ten years, the Carter County Museum has seen a renaissance led by a local kid, Nate Carroll, who went to college at the University of Southern California. Nate started inviting his Los Angeles fossil and art friends to visit Ekalaka and soon the whole town was punching above its weight. Each summer the museum hosts the Dino Shindig complete with a pitchfork barbeque, a street dance, and a full-on paleontology symposium. At a time when lots of high plains towns are drying up and blowing away, Ekalaka boasts a growing population and bright fossil-fueled future.

After visiting the museum and bumping into some old friends, Ray and I set off across the gumbo and into a dusty afternoon. The land to the south and west is a huge, empty corner of Montana, and our drive continued for hours as rolling thunderheads dumped periodic showers.

Around sunset, we realized that we were on the same path that Custer had taken to his demise at the Little Bighorn. About 8:45 P.M., with only minutes before closing, we drove into the Custer Battlefield monument. The massacre site was completely deserted except for one nervous security guard. Big Blue glided up to the top of the hill where a big concrete memorial marks the place where Custer fell. The moon was full, and we looked down the hill at the white headstones dotting the slope, each marking the spot where some poor trooper got his comeuppance in the form of a bullet, arrow, or tomahawk. Just then, a smooth, strong wind came up, and we both looked at each other and got truly spooked. We hurriedly hopped in Big Blue and headed to the little town of Ranchester, Wyoming.

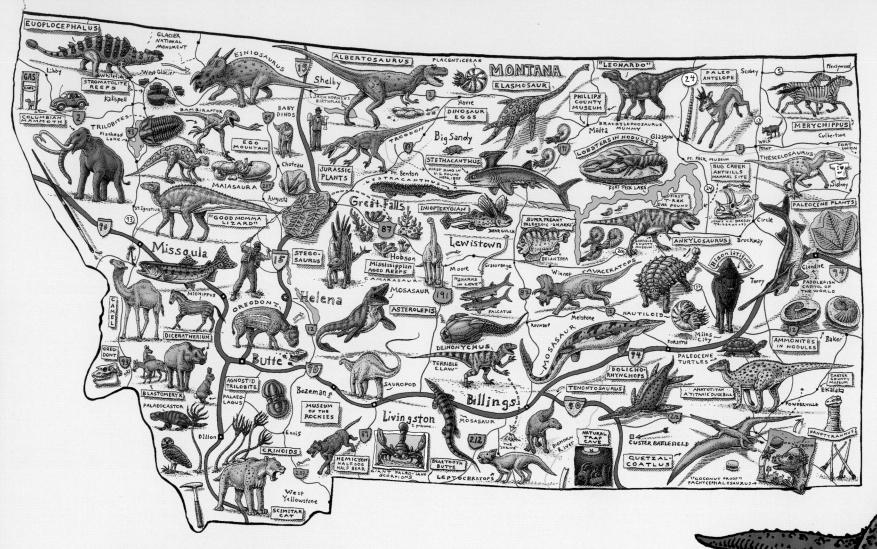

Montana Is for Bone-Diggers

Eastern Montana does not live up to the state's name. Instead, it's a huge expanse of grasslands and badlands and a high-speed drive-by for those headed for Seattle on Interstate 90. What it lacks in topography, it more than makes up for in fossils. The Yellowstone and Missouri rivers drain the distant Rockies and lazily slice their way across the eastern half of the state. In the process, they've exposed Late Cretaceous and Paleocene rocks in some of the world's most extensive and fossiliferous badlands. The town of Glendive on the Yellowstone River near the North Dakota border is best known for its riverbed agates and for hundred-pound paddlefish, but it also hosts Makoshika State Park, a hoodoo heaven dripping with *Triceratops* bones. Jack Horner made himself, and the town of Choteau, famous when he found baby dinosaur bones in a coffee can in a rock shop and traced them back to their source,

which turned out to be a dinosaur nesting ground full of nests, eggs, and babies. Eventually he also discovered bone beds composed of thousands of duckbilled dinosaurs, which he named *Maiasaura*, "the good mother lizard." In time, the state of Montana was forced to acknowledge its riches and funded the expansion of the Museum of the Rockies in Bozeman. Although it tips its hat to settlers and Native Americans, this museum is really a tribute to the star power and dinosaur-finding ability of Jack Horner.

In the forgotten center of eastern Montana, the area around Jordan has been haunted by paleontologists ever since Barnum Brown found the first *Tyrannosaurus rex* skeleton there in 1902. Legendary amateur collector Harley Garbani scored a pair of *T. rex* for Los Angeles here in the 1960s, and in the 1970s the place was overrun by hunters of tiny Cretaceous and

Paleocene mammals. In the 1980s it became ground zero for the K-Pg boundary debate, providing most of what we know about how dinosaurs and smaller animals responded to the Chicxulub impact. In the late 1990s, Horner began prospecting the southern shores of Fort Peck Reservoir (a dammed portion of the Missouri River). In the process, he discovered so many *T. rex* skeletons that he got bored by the king of dinosaurs. In 2000, Phillips County coughed up a nearly perfect mummified *Brachylophosaurus* duckbill that was named Leonardo because of some 1917 graffiti on the rock near where it lay. Leonardo is cited by the *Guinness Book of World Records* as the world's best-preserved dinosaur because more than 90 percent of its skin is intact. This world-class fossil can now be seen in the museum in the tiny town of Malta.

In Ray's mind, the fossil epicenter of Montana is located near Lewiston at a place called Bear Gulch in the Big Snowy Mountains. Discovered in 1967, this obscure 318-million-year-old pocket of platy limestone preserves an unbelievably cool equatorial seafloor. Known mainly for its insanely fine preservation, Bear Gulch has produced dozens of species of fish with intact skin, color patterns, gut contents, and even sexual organs. The pièce de résistance of this amazing site is a small slab with a pair of six-inch-long *Falcatus* sharks fossilized in the act of copulation. It is one of the rarest of fossils, an amorous couple engaged *in flagrante delicto*.

Drs. Dick Lund and Eileen Grogan have spent decades excavating the site along with the help of an army of mostly local volunteers. They've unearthed an astounding array of Paleozoic fishes, nearly all of them new to science. The Bear Gulch fauna was dominated by a wild menagerie of sharks that have long captivated Ray's artistic eye. In 1998, the Discovery Channel flew Ray out to meet Dick and Eileen live and on-camera at Bear Gulch. Since then, Ray has worked with the two to create lifelike reconstructions of their discoveries.

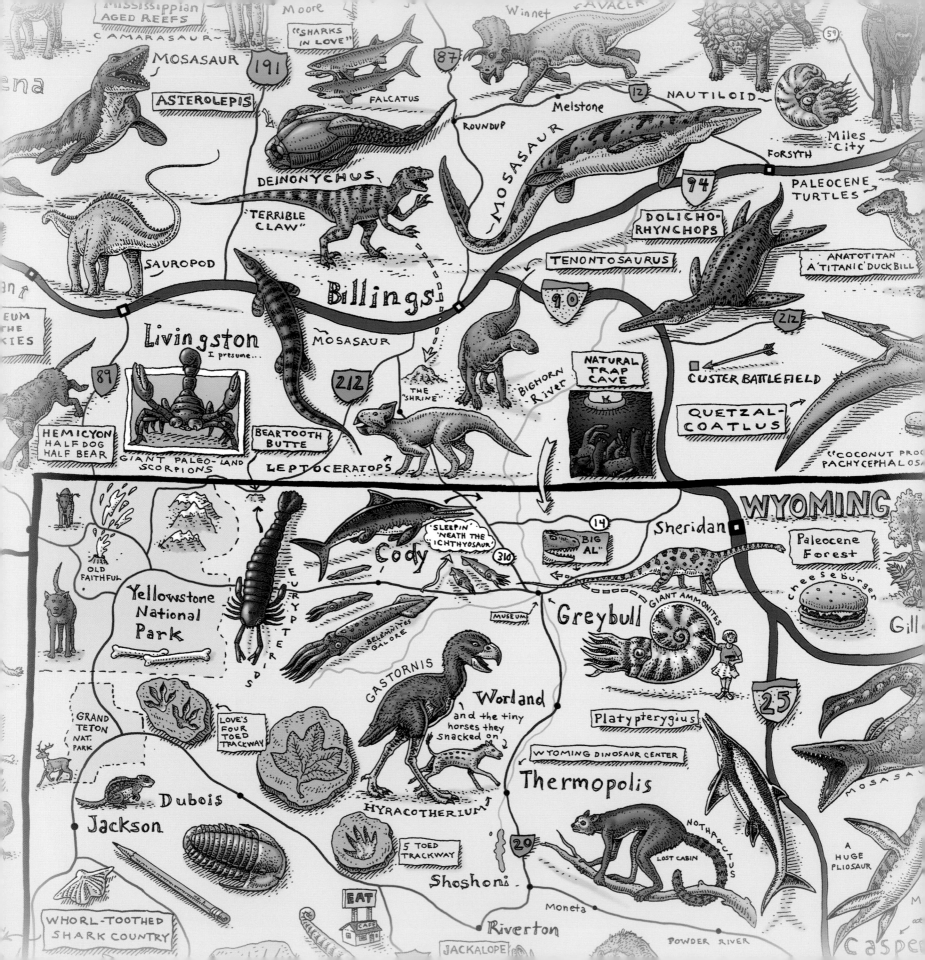

8
40,000 MAMMALS CAN BE WRONG

The top of the Bighorn Mountain range is bucolic, or at least a moosey version of bucolic, with meandering streams and green patches of mountain spruce. It was here, at the western edge of this rolling surface, that ancient Indigenous people arranged chunks of Bighorn Dolomite to make the famous Medicine Wheel. This stony pinwheel looks like a giant wagon wheel with 28 spokes. Dozens of these sacred stone circles also exist on the prairies of Montana and Alberta. The chunks of rock in this wheel contain fossils of animals that lived at the bottom of a sea 400 million years ago.

At the western edge of the planed mountaintop, the road begins its downward slide and suddenly, there is 7,000 feet of air below you. This giant hole in the ground is the Bighorn Basin. The structure is so big that you can easily see it on any map of North America that shows some topography. It's the northwest-southeast oval that dangles off the northern border of Wyoming. For paleontologists, it's also the promised land. Taken as a whole, the Bighorn Basin probably has more exposed rocks and more fossils from more formations than any other place in the world. I've often thought that if I could have only one place on Earth to tell the story of life on Earth, the Bighorn Basin would be it. The story isn't quite complete, missing most of the Precambrian, the Silurian, the Miocene, and the Pliocene. Nonetheless, the parts that are there are huge and magnificent and could make a book all by themselves.

The Bighorn Basin is dry and difficult to find. Surrounded by five mountain ranges, this huge, dry depression was hard to get to and offered little incentive for those attempting to reach it. The Bozeman Trail missed it by going north, and the Oregon Trail missed it by going south. Jacob Wortman, the first paleontologist to reach this promised land, accessed the basin from the south in 1880. His spectacular finds of Eocene mammals inspired additional attempts. Paleontologists still come to the Bighorn Basin each summer, realizing that they've barely begun to scratch the surface of this amazing place.

After stopping at the U.S. Forest Service visitor's center at Burgess Junction and admiring large, chunky Paleozoic clams and corals, Ray and I hopped into Big Blue and headed down into the basin. The geology of the western flank of the Bighorns is textbook spectacular. As the flat-topped Bighorn Mountain range rose, it was covered by more than 8,000 feet of flat-lying layers of sedimentary rocks. Along the margins of the range these same layers were tilted, folded, broken, and overturned. Today, the overlying layers are all gone but the chaos on the margin of the range remains. Driving down the steep road known as Alternate U.S. Route 14, we could see ancient seafloors now standing with perfect verticality. In places, the beds have literally been overturned. We stopped to look at this upside-down world and found 400-million-year-old marine creatures. Ray stood on his head to orient his body like an Ordovician clam.

Near the bottom of the hill, the road crosses the giant thrust fault that separated

These vertical fins of Mississippian Madison Limestone along the flanks of the Beartooth Mountains near Red Lodge, Montana, are a great example of an ancient seafloor turned on its side.

basin from mountain some 60 million years ago. Here, upside-down Paleozoic sand dunes lie atop right-side-up Triassic red beds. The rocks in the fault zone are shot through with weak spots, and the road slides on a regular basis. The fault zone is easily recognized by fresh applications of asphalt on the ever-breaking road. Once across the fault and into the basin, the layered rocks slope gently downhill. These endless exposures invited inspection, and we were soon off-road looking at the multihued rocks of the Morrison Formation.

In the Bighorn Basin, the stack of formations is more than 10,000 feet thick. Fortunately, there are several formations that are so distinctive that they orient you to where you are in the stack. The Morrison is one such formation, as it has brilliant red, white, and blue banding. Above the Morrison is a mile thick sequence of Cretaceous marine mud-

stone and sandstone layers. The mudstone weathers into gray rolling hills unencumbered with plant life. Whenever I see Cretaceous marine shale, I think of ammonites. And when I think of them, I want to find one.

We drove down a long road into an area of bentonite mines. Bentonite is an altered form of volcanic ash. All throughout the Late Cretaceous, erupting volcanoes in western Montana dusted the surface of the sea with a fine sifting of airborne ash. If the eruption was large enough, the ash would settle through the water column and form a soupy layer on the seafloor. Some bentonite layers are paper-thin, and others are as thick as a house.

Bentonite has a number of industrial uses, but it is most commonly used to lubricate drill holes for oil rigs. When bentonite at the surface is moistened by rain, it gets unbelievable slippery. Many people have learned this at their peril when driving across these old ash layers. Bentonite is moderately big business in the Bighorn Basin, and mining it's simply a matter of finding a layer and digging it out. These mines are good places to see freshly excavated blocks of old seafloor and to hunt for the remains of critters that got buried in the soup. The day was plenty hot, so there was little danger of slipping anywhere. We drove into an old bentonite mine where desk-sized blocks of old seafloor were piled into huge jumbles. Ray and I searched for fossils by crawling from one block to the next and reading the surfaces. I found a knobby ammonite the size of a dinner plate, but it had been crushed flat when the shale compressed. We found a few fish scales, and then gave up.

As we headed back down the dirt road, Ray demanded that I stop the truck. He had a feeling about the rolling hill of Cretaceous mud next to the road, and he needed to respect the feeling. I couldn't see any obvious concretions, and the site looked barren to me, but you can't find fossils if you don't look, so I stopped the truck. Ray grabbed a hammer and strolled off into the heat. By the time I'd begun to lace my boots, Ray was yelling that he'd found a plesiosaur. I ran up the popcorn slope, filling my open boots with Cretaceous crumbs. Ray was holding a beautiful three-inch fossil bone, a vertebra for sure, most likely from a plesiosaur, and he was grinning from ear to ear. We crawled around looking for more, but that was all she wrote. It was Ray's first "find based on a feeling," and he was pretty proud of himself. Being a vertebrate fossil on federal land, the specimen was not destined for Ray's pocket. But this time we did have a collecting permit, so I logged the vertebra into the museum's record book.

Continuing down the hill, we drove to the banks of the Bighorn River. This is a river that respects no mountain ranges, cutting through both the Owl Creek and Bighorn Mountains on the way to its confluence with the Yellowstone River. Like many rivers in the Rockies, the Bighorn is literally older than the surrounding mountains. Not just older than the hills, but older than the mountains. Evidence for this bold claim is plain to see. The headwaters of the Bighorn River start at Togwotee Pass, the divide between the Wind River Basin and Jackson Hole (this is also the Continental Divide). Here it's called the Wind River. It flows southeast to Shoshoni, Wyoming, where it bends sharply north and flows directly into the Owl Creek Mountains through Wind River Canyon. At the north end of the canyon, the river abruptly changes its name to the Bighorn

at a place oddly named the Wedding of the Waters and flows into the Bighorn Basin. Usually the wedding of the waters is where two rivers flow together, but here a single river just changes its name. The Bighorn River performs a similar trick at the north end of the basin, slicing through the Bighorn Mountains and forming Bighorn Canyon. The only sensible explanation for this weird pattern is that the river is older than the mountains, the canyons, and the basins.

We crossed the river and headed into Lovell to visit the forest service visitor's center, which has a great relief map of the northern margin of the basin. By this time, the pleasant morning had morphed into one of those smoking hot Wyoming afternoons that melts the ice in your cooler and makes you want to retreat to the nearest dark bar. Instead, we headed back east, recrossed the Bighorn River, and drove north on a dirt road to find a legendary deep and deadly hole in the ground.

The road to Natural Trap is one of those roads that you drive not really knowing if it's drivable. After consulting a freshly purchased topographic map, we started up a steep valley. The road is made of solid rock that got narrower, steeper, and more rutted as Big Blue crept along. Eventually, I was hanging on to the door to keep from sliding across the bench seat, and Ray was imploring me to turn around, a maneuver that seemed utterly impossible and worse than carrying on and hoping for the best. In compound low, Big Blue was up to the task, and we emerged from the rocky canyon onto a sage-covered plateau that stretched to the north. After a few miles, we pulled up to the gaping mouth of a cave that has killed more than 40,000 animals in the last 100,000 years.

Troll at the grated mouth of Natural Trap Cave.

FORTY THOUSAND MAMMALS

CAN BE WRONG

Natural Trap Cave is shaped like a short-necked beer bottle. The opening, roughly 20 feet wide, appears as an ominous hole in the floor of a gentle valley on the top of a limestone plateau. The BLM has built a metal frame over the cave's gaping mouth, so we were able to walk out onto it and peer down into the inky blackness. Even though the spacing of the metal bars is so close that it would have been impossible to slip through, the fact that this cave was the site of so much fatality gave us pause as we considered its victims.

Natural Trap is the greatest pit trap of all time. Lying at the bottom of an 85-foot drop are the skeletons of mammoths, horses, pronghorns, bighorn sheep, camels, musk oxen, bison, fox, wolverines, dire wolves, coyotes, gray wolves, red fox, lynx, weasels, American lions, short-faced bears, and American cheetahs. Each one of these animals made a bad decision and fell into the hole. The drop was enough to kill them. Regardless, there was no way out. There was no way in for big-bodied scavengers either, so the bodies lay where they fell, essentially undisturbed. In time the floor of the cave was covered by a pyramid-like pile of fallen skeletons.

In order to study the site, Larry Martin and his crew from the University of Kansas built a staircase on scaffolding and climbed down into the hole. Each year from 1974 until 1985, they dug away at the pyramid pile of carcasses on the cave floor. Results of their digging show the census of bad luck. The University of Kansas Museum in Lawrence, Kansas, displays nearly perfect skeletons of a giant musk oxen (*Bootherium bombifrons*), an extinct bighorn sheep (*Ovis catclawensis*), and an American cheetah (*Miracinonyx trumani*) from the cave. Peering down into the dark hole, we looked at each other and smiled. Here was one of the best fossil sites in the world.

On the way back to town, we stopped at the mouth of Cottonwood Canyon. The Bighorns are a splendid example of Dave Love's metaphorical pigs waking up under a stony blanket. Thick Paleozoic limestone and dolomite layers are draped across the mountain margin. Deep canyons slice these layers and create a series of flatirons shaped like piranha teeth. Cottonwood Canyon is one of these canyons, a place that would be a national park if it were located in a less-endowed state. In Wyoming, it's just another spectacular canyon with a crappy gravel road that goes nowhere. Its walls are nearly 2,100 feet high. The canyon wall exposes an amazing stack of marine rocks from the Cambrian, Ordovician, Devonian, and Mississippian periods.

The Beartooth Butte Formation, found in places along the walls of Cottonwood Canyon, is a rare one, occurring only in a few places in Wyoming. The formation formed when a Silurian or Devonian river cut canyons into the top of the Ordovician dolomite. This happened during a brief interlude when the sea drained off the continent and old seafloors were high and dry. In the Early Devonian, the sea level came back up and the canyons were drowned with water, creating fjords that filled with mud. Today, this mud is preserved as lenses of red mudstone at the top of the Ordovician dolomite. These layers have fossils of some of the first fish, some of the first land plants, and some terrifying arthropods.

The potential of the Beartooth Butte Formation was discovered in 1931 by Princeton professor Erling Dorf, who led a team of students into the high Beartooth Mountains near Cooke City, Montana. His team had been alerted to the presence of "fossil butterflies" by local hunters, and they set out to inspect the report. The butterflies turned out to be the tails of trilobites in the Cambrian Park Shale. After correctly identifying them, Erling saw the red lens on the face of Beartooth Butte and climbed up to inspect it. He found a trove of twiglike land plants and the bony head plates of early fishes. At the time, these land plants were some of the earliest known in the world, and Erling acquired some fame for finding them. Meanwhile, the game warden in Red Lodge heard rumors of Princeton students fishing without licenses in Beartooth Lake and set out to bag himself a bunch of eastern scofflaws. When he rode into Erling's camp and learned that the fish were not fresh trout but 395-million-year-old Devonian fossils, he gamely gave up and went home.

The exposure of the Beartooth Butte Formation in Cottonwood Canyon was discovered later. I had a hunch that this would be a great place to reconstruct a Devonian landscape and led a team to this site in 1993. We camped at an old homesteader cabin at the mouth of the canyon and hiked into the site every morning. We were only a quarter mile from the quarry, but we were 1,000 feet lower, and each morning started with a grueling hour-and-a-half climb.

The mouth of Cottonwood Canyon, home of giant scorpions and eurypterids.

He lay especially still, hoping the eurypterid wouldn't notice him.

Our team of seven included the Patagonian field man Pablo Puerta, renowned on five continents for his ability to find amazing fossils. Pablo held the record for fastest trip from camp to quarry with a smoking 42-minute time, while none of the rest of us could even crack the one-hour mark. It was all the worse because he had enough energy and breath to sing as he climbed.

The quarry site had a magnificent view down the canyon and out into the Bighorn Basin. Digging was straightforward, since the rock wasn't covered by much dirt. We just pried out blocks and split them with hammers and chisels. This site had abundant remains of early land plants and even some unmistakable fossil roots, some of the earliest in the world. A Field Museum excavation in the 1960s had yielded a complete skeleton of a bony-headed fish, and there were rumors of large fossil euryp-terids, feared marine predators related to spiders and scorpions. We had luck finding chunks of bony-headed fish, but little else. On day three of the dig, I drove into town for supplies, and when I returned, the quarry was abuzz. Two of the team had found a giant claw. Unfortunately, the portion of the claw that connected to the body was oriented away from the cliff, meaning that the body had long ago weathered away. When I got a close look at the big chomper, I realized that I was looking at a piece of a eurypterid known as *Pterygotus* that must have been at least five feet long. I later learned that this site has also produced a rare land-roaming scorpion known as *Phaearcturus*, which was almost three feet long.

Three hundred and ninety-five million years ago, Wyoming lay south of the equator and was covered by a tropical sea. As those waters receded, streams flowing from the land carved shallow valleys onto exposed wave-cut platforms. The land would have looked barren at a distance, but closer inspection would have shown patches of plants, some as tall as a few feet. Wandering around in this miniature canopy were truly massive and murderous arthropods. The scorpions were bigger than badgers, and the eurypterids may have been amphibious. This was our world before our world even began to look vaguely familiar.

I explained all of this to Ray as we stood at the mouth of the canyon watching a glorious sunset light up the layers. Thoughts of enormous scorpions crawling into his bed sent Ray scurrying for his notebook. He wanted to climb into the land of the eurypterid, but I was keenly aware of the effort involved and tried to sate him with stories instead of firsthand experi-ence. After horsing around and taking some pictures, with Ray quickly sketching giant murderous sea scorpions and armored fish swimming above the canyon walls, we set out from the sunset toward the appropriately named Sundance Formation.

The Sundance Shale is perhaps the most fossiliferous formation in Wyoming. This gray mudstone was once the mud at the bottom of a Jurassic sea full of oysters, shelled squid, and sleek, dolphin-shaped ichthyosaurs. Today, the shale weathers to form slopes and valley floors. Lying below the dinosaur-rich Morrison and above the distinc-tive bright-red Chugwater Formation, the Sundance is the easiest place to find a fossil, or your money back. In many places, the ground is littered with fossils. Bullet-shaped belemnites and ram's horn *Gryphaea* oysters are the most common finds, but with a bit of patience you can lay your hands on star-shaped crinoid stems and bits of marine lizard. On the northern end of the Bighorn Basin, the Sundance Formation includes a layer of paper shale, the remains of some dried-up coastal lagoon. The fossil fish our Dinopalooza friends Chris and Dave were mining on their nearby ranch had come from this layer.

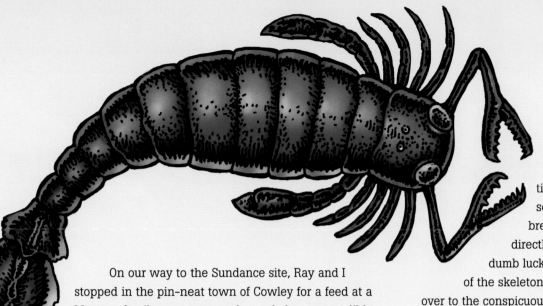

On our way to the Sundance site, Ray and I stopped in the pin-neat town of Cowley for a feed at a Mormon family restaurant and sampled a very credible chicken-fried steak. By the time we finished, the sun had set over our bottomless iced teas. Chris and Dave had given us good instructions to their site, but there's an art to finding a place out in the sage at night. We pushed on into the dark, and I followed the instructions as best I could, meandering for miles down dirt roads. After an hour of tentative creeping, we arrived at a spot that I figured was close enough, and we stopped the truck to pitch the Troll family tent next to the road. It was only our second camp on this outing, and we soon discovered that the tent stakes had been lost in the post-Dinopalooza hangover haze or, if on board, were too well hidden to be found. Not a problem, as the capacious toolboxes of Big Blue yielded a profusion of pickaxe heads and crowbars that soon pegged the tent to the soft soil. We tucked into our bags and began chatting about Jurassic seas. Ray was completely jazzed and couldn't stop talking about big-eyed, sleek marine lizards, wondering out loud if they breached like dolphins.

When morning came, I set about building a pot of coffee on the back of the truck. A bleary-eyed Troll poked his head out of the tent and complained that the ground was too rocky for a good night's sleep. We had pitched our tent on the Sundance Shale, and the bullet-shaped bumps under our bedrolls were fossils. Troll started collecting them while still in his sleeping bag, and soon he had a handful of belemnites. We sipped our coffee and read Chris's notes about the buried ichthyosaur. The instruc-tions were good, and we had followed them well.

The modern world of paleon-tology is armed with GPS devices, so there's no excuse for bad direc-tions. With Chris's coordinates, we soon unburied his marine lizard for a breakfast viewing high up on a slope directly above the Troll family tent. By dumb luck, we had camped within 50 yards of the skeleton. We reburied the find and ambled over to the conspicuous outcrops of the Hulett Paper Shale to find some fish.

Paper shale sometimes lives up to its name, and the Hulett is amazing in that way. It's possible to pull out book-sized and book-shaped chunks and then split off the pages one at a time. The sheets are stable and sturdy, even when they're only a few millimeters thick. Using sharp knives, we sliced open pages of time and looked at what had been buried in a Jurassic lagoon when Wyo-ming was under the sea. In no time, we were finding fish and insects.

The concept of Middle Jurassic paper shale from a marine lagoon is particularly tantalizing when you stop to consider that one of the most famous fossils ever found came from Late Jurassic lagoonal deposits in Bavaria. *Archaeopteryx*, the first feathered dinosaur/earliest bird, was discovered in 1860, the year after Darwin published *On the Origin of Species,* and provided an exquisite example of a fossil animal with characteristics of two major animal groups: birds and dinosaurs. More than a century later, Bavaria's Solnhofen Limestone has coughed up only a dozen more *Archaeopteryx* skeletons. Mean-while, slightly younger paper shale from Liaoning Province in northeastern China has produced a whole flock of fossil birds and many feathered dinosaurs. What we don't yet have is a really credible feathered dinosaur/bird earlier than the Late Jurassic.

The same Larry Martin who revealed the joys of Natural Trap Cave has spent months splitting the paper shale of the Sundance Formation in the hope of finding

Dialing in an ichthyosaur with a GPS. The Troll family tent and Big Blue are located at the left end of the road.

such a creature. That's the way predictive paleontology works. Get a question, in this case, "What is the oldest bird?" then choose a strategy that might allow you to find it. The Sundance is a great place to look, though, and I am pretty sure that Larry's bird will show up one day.

Another part of the *Archaeopteryx* story played out not far to the west of where we sat. The Cloverly Formation is an Early Cretaceous pile of sandstone and mudstone that's exposed several hundred feet up the section from the Sundance. The American Museum's ubiquitous Barnum Brown prospected the Cloverly late in his career and found several provocative dinosaur fragments that he never got around to describing. In the early 1960s, Yale's John Ostrom used Brown's field notes and worked his way around the southern end of the Pryor Mountains near Bridger, Montana, in search of new and interesting dinosaurs. On the last day of the field season in 1964, Ostrom found a skeleton of an animal that he later named *Deinonychus.* The deadly grace and implied pack behavior of this wolf-sized meat eater was yet another argument for the warm blood of dinosaurs and the dinosaur-bird link. *Deinonychus,* under the pseudonym of its cousin *Velociraptor,* would go on to be the star of *Jurassic Park.*

Though fish and insect fossils lay in sheets at our feet, we got bored with not finding any Jurassic birds, so we stopped splitting shale and drove to Greybull. This

(left) Sleeping with the ichthyosaurs.

Pointing at Sundance belemnites.

sleepy eastern basin town has a great local museum that displays some of the amazing fossil richness of the Bighorn Basin. My favorite fossil there is a giant ammonite collected from the nearby Cretaceous marine shale. Almost five feet in diameter, this giant *Parapuzosia bradyi* is one of the largest ammonites ever collected in North America. It was collected in 1963 in the Cody Shale and there must be more of them out there. The world record goes to a German *Parapuzosia*, which stretches the tape measure to an awesome six feet.

The road east of Greybull heads back toward the mighty Bighorn massif, but we turned off to the north and wound our way back down through the gray Cretaceous shale and into the multihued Morrison. It was in this valley that one of the greatest dinosaur hunters of all time, Barnum Brown, made one of his biggest finds. Brown was told of the site by a rancher in 1932 and led a massive excavation there in 1934. The early photographs of Howe Quarry show an unbelievable jumble of bones. Paleontologists make detailed maps of quarries that show the position of every bone. The quarry map from the Howe quarry looks like a massive dinosaur log jam. The rearing *Barosaurus* that graces the entrance of the American Museum of Natural History was one of the many fine skeletons that came from this site. The Sinclair Oil Company underwrote this expedition for the museum, and the famous green Sinclair dinosaur was a result of this collaboration. It's because of this history that many people today hold tenaciously to the misconception that oil is made from dinosaurs.

One of the common happenstances of the modern West is that commercial dinosaur diggers have realized that many of the classic quarries of the last century, such as Howe Quarry, were not dug to exhaustion. If the quarry sites can be relocated and if they're on private ranch land, then it's legal and often possible to make an arrangement with the rancher for the site to be reopened. In 1991, Kirby Siber, a Swiss dinosaur enthusiast, located the forgotten Howe Quarry and obtained a lease from the landowner to reopen it.

What was good for Barnum was also good for Kirby. The quarry was far from kicked, and bones once again began to roll out of the hill. Kirby's operation hit a snag when a BLM pilot spotted a new road near the old Howe Quarry and reported it to the BLM office in Worland. On the ground, Kirby and his team had located a splendid and unusually large *Allosaurus* skeleton with a nearly perfect skull. His mounting excitement morphed to utter dismay when BLM officials showed up and surveyed the site. Kirby was digging on the correct side of the fence, but the rancher had mistakenly set his fence on BLM property. Kirby walked away from the *Allosaurus* and back onto private land. The BLM looked around for a paleontologist who would finish the excavation and curate the specimen at a federal repository. Eventually, Jack Horner and his crew showed up from Bozeman, Montana, and finished pulling "Big Al" out of the ground. Now Al's skull graces the cover of BLM's fliers about fossil collecting on federal land, Al's skeleton resides in Montana, and Wyoming paleontologists are still sore that yet another Wyoming dinosaur has left the state.

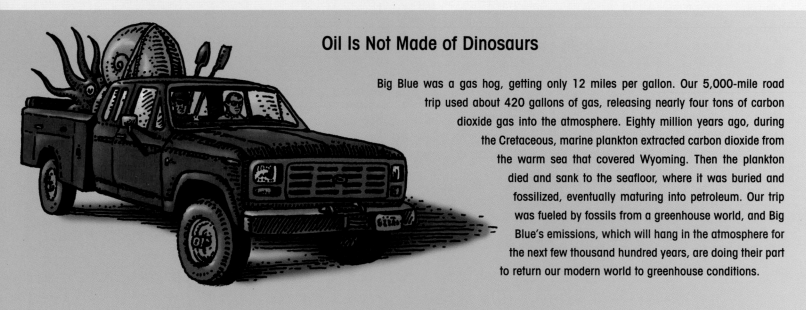

Oil Is Not Made of Dinosaurs

Big Blue was a gas hog, getting only 12 miles per gallon. Our 5,000-mile road trip used about 420 gallons of gas, releasing nearly four tons of carbon dioxide gas into the atmosphere. Eighty million years ago, during the Cretaceous, marine plankton extracted carbon dioxide from the warm sea that covered Wyoming. Then the plankton died and sank to the seafloor, where it was buried and fossilized, eventually maturing into petroleum. Our trip was fueled by fossils from a greenhouse world, and Big Blue's emissions, which will hang in the atmosphere for the next few thousand hundred years, are doing their part to return our modern world to greenhouse conditions.

Kirby learned his lesson about good fences making good neighbors and had the rest of the private ranch carefully surveyed. Luckily for him, Barnum's bone bed is huge and a whole lot of dinosaurs lie beneath private soil on the rancher's side of the fence. Since they walked away from Big Al, Kirby's team has continued to hit it big, collecting skeletons of *Apatosaurus, Camarasaurus, Dryosaurus, Stegosaurus*, and not one but seven *Diplodocus*. In 2003 they found Big Al II, an even bigger version of the one that got away. They've also found some mystery fossils: teeth of a very large unknown theropod and an extremely large rib bone for an extra-jumbo sauropod. The Morrison Formation is like that: chock-full of common dinosaurs, with the occasional tantalizing fragment of some completely unknown beast.

It was a hot afternoon when Ray and I nosed Big Blue into Kirby's camp. At first it appeared that nobody was around, but then we realized that the Swiss take siestas, a sensible strategy in the blazing-hot Bighorn Basin, and the whole crew was conked out in their tents. We eventually found a lone insomniac washing his hair. He woke Kirby and we got a tour of a new *Camarasaurus* quarry. It was an unlucky site: the digging was easy because of the thin overburden, but the fossil bones were being trampled by grazing cattle because they were so shallowly buried. As the team dug the site, they found evidence that the bones had also been trampled by fellow dinosaurs back in the Jurassic. This was a twice-trampled quarry, a recipe for well-crushed bones.

Many of Kirby's dinosaurs ended up in Switzerland, where they can be seen at the Aathal Dinosaur Museum. One of the more unexpected realities of modern dinosaur hunting is the sheer number of skeletons discovered and collected each summer. Many of these specimens end up in American museums, but a large number make their way to museums in other countries. The United States is the world's largest exporter of dinosaur skeletons because the other big dinosaur-producing countries (Mongolia, China, Canada, and Argentina) have laws that prohibit the export of dinosaurs.

We left Kirby as the sun was setting. The next morning we drove into the center of the Bighorn Basin, where huge expanses of badlands have formed in a red-and-white striped formation known as the Willwood. This and the underlying Fort Union Formation span the time from just after the extinction of the dinosaurs through the time of the superwarm climate of the Early Eocene, or from about 66 until 52 million years ago. This is when mammal survivors of the giant K-Pg cataclysm diversified and formed the major mammalian lineages that are with us today. These Bighorn Basin fossils aren't much to look at—tiny teeth and jaw fragments—but they're the world's best record of this key time in Earth history.

The Eocene mammals of the Bighorn Basin were first discovered by Jacob Wortman in 1880. Wortman returned in the 1890s, and soon others followed. A doctor in Bear Creek, Montana (a place now best known for a saloon that hosts pig races in the summer and iguana races in the winter), found some teeth in the local coal mine and sent the specimens back East for inspection. In 1929, Glenn Jepsen, a Princeton professor, began to work the area north of Powell, Wyoming, and discovered several key Paleocene mammal quarries. One, the Mantua Lentil, contained bones of small mammals that lived immediately after the dinosaur extinction, and it provided some of the very best early views of the world of survivors. The Mantua Lentil is a giant fossil streambed frozen as a series of sandstone cliffs. Jepsen's teams dynamited the cliffs and picked through the pulverized boulders for tiny jaws. Jepsen's successors still work the basin center each summer, crawling the ground for tiny teeth and jaws.

Paleomammalogists are a focused bunch. Many of the fossils they seek are literally as small as a pinhead, and you don't find a fossil pinhead by walking around: you assume the position, with nose and knees to the ground and butt high in the air. Each summer, paleomammal teams hailing from around the nation descend on the Bighorn Basin to find new fossils. They've been doing this for generations and have made close friends with the people who live in the basin. The Churchill family, broccoli farmers from Powell, Wyoming, have been hosting paleontologists for three generations. Each summer, their Fourth of July pig roast sees dozens of paleontologists who have abandoned their camps and made the pilgrimage to the Churchills' backyard to trade stories and show off their recent finds.

Fossil mammals are vanishingly rare in the Paleocene rocks but become easier to find in the Eocene. The Paleocene-Eocene boundary was marked by an extremely warm spell in the world's climate know as the Paleocene-Eocene Thermal Maximum (PETM). The reasons for this abrupt warming appear to be related to a massive release of carbon dioxide in the North Atlantic due to volcanism. The Late Paleocene was already a time when no polar ice existed anywhere in the world and the majority of the planet was forested, so the additional warming pushed subtropical climates as far north as the Arctic Circle. Because of its similarity to our present carbon dioxide crisis, the PETM in the Bighorn Basin has become a hotbed for climate warming research.

The layers of rock in the Bighorn Basin that were deposited during the P-E boundary interval are amazing in that they contain a bizarre fauna of dwarf mammals.

You may have heard of the dawn horse *Eohippus* (also known as *Hyracotherium*) and know that it's the four-toed ancestral horse often compared in size to a cocker spaniel. Amazingly, the *Eohippus* of the P-E boundary interval are dwarfed. It's kind of like saying you have a dwarfed Shetland pony. Take something small and make it smaller. It's not clear what aspect of global warmth caused small mammals to get smaller, but when the intense warming let up, the mammals started getting bigger right away.

The diversity of mammals in the Early Eocene is stunning. The most common find is a beast named *Hyopsodus*. Probably about 50 percent of the jaws found when you dig into Early Eocene formations belong to this ancient artiodactyl. When I asked paleomammalogist Richard Stucky to describe this animal to me, he said that it was sort of a smallish, cylindrical, sheeplike animal, and thus the name "tube sheep" was born. Turtles are also common. Large animals are rare and come in one of two flavors: the hippolike, tusked *Coryphodon* and the truly bizarre giant bird known as *Gastornis*. This mighty

Paleomammalogists on the prowl.

fowl stood about six feet tall, had a lower jaw that was 14 inches long, and had stubby little nubs for wings. Known as the "terror crane," it has long been thought of as the successor bully to the dinosaurs. More-recent speculations have suggested that this bird was more of a mellow, nut-cracking chicken than a predatory danger to mammals.

Hyopsodus, the "tube sheep."

Gastornis dining on a dawn horse.

Almost everything else is pretty small. Mammals started small, and many of the lineages stayed small as they diversified. The Willwood Formation has a mixture of mammals from living groups such as rodents, primates, bats, insectivores, artiodactyls, perissodactyls, and carnivores, and from a smattering of extinct groups such as mesonychids, condylarths, creodonts, taeniodonts, and others with even odder names. Phil Gingerich from the University of Michigan has been returning to Wyoming since the early 1960s, and his crews have collected tens of thousands of these tiny jaws and bones from hundreds of fossil localities. The result is the most densely sampled records of mammal evolution on Earth. Through Phil's data, it's possible to see a clear example of the gradual process of evolution. Creationist apologists often claim that the fossil record is too episodic to show evolution. Phil has shown that the episodic appearance is an artifact of small sample size, not any reality of the fossil record.

While most of these fossils are fragmentary, diligent efforts by paleontologists have slowly assembled more-complete skeletons. In paleospeak, anything behind the skull is postcranial, and for most Paleocene and Early

Eocene mammals, we know very little about their postcranial skeletons. For the last 40 years, a team from Johns Hopkins led by Ken Rose has been carefully collecting and assembling postcranial skeletons of fragmentary Eocene animals.

More recently, two students from the University of Michigan have found some truly spectacular fossil skeletons by dissolving blocks of limestone in weak acids. The limestone apparently formed in shallow ponds or depressions in Eocene forests. Somehow, the complete skeletons of exquisite tiny bats, primates, and insectivores have been completely preserved in these tiny, limy tombs.

The Bighorn Basin is likely the most-studied site in the world for the topic of mammal evolution, but this information is not that well known in the basin itself. The local lore and tourist fodder trends more toward cowboys and Indians, Buffalo Bill, and Yellowstone. The town of Cody has been a tourist trap ever since Buffalo Bill opened the Irma Hotel, "just the sweetest hotel that ever was," on Main Street in 1902, just 22 years after Wortman discovered the adjacent fossil fields. Every summer, millions of tourists roll through town, sample the daily rodeo, and roll on to Yellowstone, ignorant of the fact that some of the world's best fossil sites are right there.

Here we were, on this great trip through space and geologic time, seeing dozens of fossil sites every day, and everybody else on the road couldn't care less. We ducked into the palatial Buffalo Bill Museum, which had just added an entire wing, the Draper Museum of Natural History, dedicated to local nature. The focus of the museum was all ecology and no evolution. Didn't

the designers of this new museum know that they were located in the best fossil basin in the world? Didn't they care that the living ecosystems of Wyoming are the result of billions of years of evolutionary change? We realized that ours was a simple quest. We just want people to know that the Earth has a long and exciting history and that life has persevered despite asteroids, glaciers, and volcanoes. The evidence for this vast drama is all around us: fossils are everywhere. In 2010, the town of Worland filled this gap by opening the Washakie Museum and Cultural Center. One of its core exhibits is an overview of the geology and paleontology of the Bighorn Basin.

Just to the north of Cody is one of the most unlikely mountains in the world. It's a mountain loved by creationists and much discussed by geologists. Geologically, it's a true conundrum. The jagged mountaintop consists of 350-million-year-old marine Madison Limestone, while the lower slopes are composed of the 55-million-year-old Early Eocene *Eohippus*-bearing Willwood Formation. Heart Mountain appears to defy the defining principle of stratigraphy, the one that says the oldest layers are found on the bottom. In its case, the oldest layers are clearly on top. For the "short-chronology" creationists who believe that the world is about 6,000 years old, this is a place where an exception to the rule tosses out the whole conceptual structure. To geologists, this site is endlessly interesting and not a little puzzling.

Detailed fieldwork now suggests that the mountain is the result of a giant horizontal fault and that the old rock of the mountaintop slid or was shoved into place sometime after 50 million years ago. The source of the mountain top has been located near the northeast corner of Yellowstone National Park, some 50 miles away. What geologists can't agree on is how long it took to travel those 50 miles. One group of scientists thinks that the mountain was pushed into place over a few million years as the Absaroka Mountains were forming, but another argues that the mountains slid into place in one massive, catastrophic landslide. Both ideas have good arguments and passionate proponents. I kind of like the idea of a mountain sliding 50 miles into place in a few moments. Talk about a change of scenery.

(left) Badlands of the Paleocene-Eocene boundary at the nose of Polecat Bench near Powell, Wyoming.

North of Heart Mountain and south of the Montana line is one of the most rewarding places in North America to look at geology. Clark's Fork Canyon is a gaping slice into the eastern margin of the Beartooth Mountains. Here, tipped on edge for perfect viewing, are nearly 10,000 feet of layered rock: a record of Earth from the Cambrian to the Paleocene, a time span of nearly 500 million years. This entire stack is piled on top of a massive 1,200-foot wall of 2.7-billion-year-old metamorphic rock. Or to put it another way, it is possible in this one spot to see rocks that that span more than half of Earth's entire 4.567-billion-year history. The road up the canyon is massively overbuilt, a highway to nowhere. The original plan was to carve an access road to Yellowstone, but the realities of the canyon won out, and now the pavement stops a few miles inside it. Like many places in Wyoming, this world-class geological hot spot isn't labeled on maps or marked by signs. It's just there if you know what to look for. This is the subtle magic of the Cowboy State.

Ray and I continued our trip down the basin, aiming for the hot springs at Thermopolis. As Big Blue roared along through this open, endless, fossil-filled landscape, my only thought was, "How could I ever explore all of this country?" My sister, a documentary filmmaker who travels the world from Brooklyn, is baffled by the ease with which I travel through space and time without becom-

ing overwhelmed by the immensity of it all. For me, it's the immensity that makes it interesting, because I know that we'll never find the last fossil. And the immensity is not overwhelming to me because of the soothing spatial framework of geologic maps and the temporal comfort of geologic time. Yes, the Earth is 4.567 billion years old, but nonmicroscopic life-forms have only been around for 600 million years or so. I'm comfortable because I can map the vastness of paleontology and fit it comfortably into my planet's timetable.

As we drove through the dark past Worland, I told Ray about all the cool digs we were passing. The basin is so rich that nearly every ridge has a story. For example, back in 1990, Scott Wing, a Smithsonian paleobotanist and my academic brother (we shared the same Ph.D. adviser, Leo Hickey), was working on Big Cedar Ridge in the southeast corner of the basin on a godforsaken piece of real estate known as the Honeycombs, a huge roadless area of brilliantly colored badlands composed of Paleocene and Eocene rocks. The sheer vastness of the Honeycombs had crushed the spirit of many an ambitious graduate student, but Scott has spent his entire career tracking the fossil plants of the Willwood Formation. He's not easily deterred by insane temperatures, long roads, and no water. He's keenly interested in what happened to plants at the Paleocene-Eocene transition, so he found himself wandering around the Honeycombs with shovel in hand, trying to find layers of fossil leaves.

It's simple geology that walking downhill in an area of flat-lying strata means you're going back in time, and that climbing the same hill means the rocks under your feet are getting younger as you ascend. For me, this takes the sting out of climbing a hill, because I know that I can climb rocks a lot faster than it took to get them there in the first place. Assuming a reasonable sediment accumulation rate of 100 meters in one million years, simple math shows that you make or lose about 3,300 years for every foot of elevation that you gain or lose. A key point here is that sediment doesn't always accumulate at a steady rate. Sometimes you can cross a horizon where time is missing (this happens either when sediment stops piling up or when is eroded away), and sometimes that amount of time can be significant.

Geologists and paleontologists like to know how old the rocks are beneath their feet. I know that I'm always aware of this obscure information. Once I had oriented Ray, he knew that he could ask me at any point along the road and I would have a ready estimate of the age of the rock below the truck.

Back in the Honeycombs in 1990, Scott thought he knew where he was, but he didn't. He'd been working in 60-million-year-old rocks and had walked down a hill thinking that he was walking into 61-million-year-old rocks. In reality, he stepped over a time gap and ended up in 72-million-year-old rocks. No big deal; it could happen to anyone. But then he stuck his shovel into a weathered gumbo slope and flipped out an amazing fossil. Still not fully aware that he'd slipped out of the Paleocene and into the Cretaceous, he began to dig ferociously and realized that the whole ridgeline was made of fossil plants. He knew that he was digging into a rock that formed from volcanic ash, and he soon realized that the fossils were at the bottom of the ash layer. As he dug, he began to realize that he was finding whole fossil plants: complete palmetto plants, complete fern fronds, darling little cycads. He had found a buried landscape, a plant Pompeii. Later, when he got the samples back to Washington, D.C., he realized that they were plants unknown in the Paleocene, and when he sent a sample of the ash to be dated, he learned that the site was 72 million years old and his fossils were from the Late Cretaceous Meeteetse Formation. Armed with this new information, he returned the next year with a big team and opened more than a hundred quarries, literally mapping a vegetational transect, or strip, through a Late

Paleobotanists Leo Hickey (left) and Scott Wing (right) at the edge of the Honeycombs near Worland, Wyoming.

Cretaceous landscape. It was the first time anybody had found such a site and followed the discovery with the absurd amount of work necessary to really reconstitute an ancient meadow. This kind of paleontology is a kind of time travel.

Today, the BLM office in Worland allows, even encourages, people to visit this site and to collect their own fossils. Unfortunately for Ray, it was another frustrating flyby, a cool fossil spot that we couldn't visit because we had other places to go.

Instead, we rolled into Thermopolis, and the next morning we enjoyed Star Plunge, a very campy spigot for the world's largest hot springs. The geology of the nearby Owl Creek Mountains is such that rain falling on the mountains sinks into the ground and percolates to great depths, where the Earth's heat warms it up to over 100 degrees. Then a great fold in the rock layers beneath Thermopolis brings it right back to the surface, where it spews out at the rate of more than a million gallons a day. Smelling like rotten eggs, it's just the thing for the aching joints of

It takes a paleobotanist to understand what the biggest herbivores of all time ate.

a pair of road trippers. Taking the waters at Thermopolis is mandatory for all travelers, and the town has been a destination since mountain men soaked their stinky bones back in the 1830s.

In 1992, a wandering German fossil enthusiast-veterinarian named Burkhart Pohl was poking around for fossils in the Morrison Formation south of the Owl Creek Mountains. A rancher told him about similar outcrops in the hills behind Thermopolis. Burkhart could read a geological map better than the average Joe and learned that the same fold in the rock that brought the hot water to the surface also brought the thick and fossiliferous Morrison Formation to the surface. The rancher's relative was a realtor and, before he knew it, Burkhart found a ranch for sale for the right price and bought it for its dinosaur potential. Sure enough, the Morrison Formation produced, and Burkhart had a dinosaur quarry going by 1993.

People who like fossils often collect them. They also like museums and often have dreams of building them. Burkhart had a huge collection of European fossils and his own dinosaur quarry, so he decided to build his own museum. The Thermopolis city fathers were happy to entertain the idea of another tourist draw, and soon enough Burkhart built a giant square building. In 1995, he opened the Wyoming Dinosaur Center, another chapter in the great American dream (isolated paleonerd version). It seems it takes a German to keep Wyoming dinosaurs in Wyoming.

It's a big museum, full of wonderful European fossils and Jurassic dinosaurs from the ranch. There are daily tours to the dinosaur mines, where volunteers and employees are excavating the back side of a hill and finding tons of bones. Some of the bones go into the museum, but others have been sent overseas. The big *Apatosaurus* at the National Science Museum in Tokyo came from Burkhart's ranch. The quarry itself is a scary affair, since the layers of the Morrison Formation tilt into the hill and the huge sandstone blocks of the Cloverly Formation loom overhead. It seemed to me that the more Burkhart digs, the better his chances are of oversteepening the slope and bringing huge cubes of sandstone tumbling down on his dinosaur diggers. Not a place for a rainy day. We headed for the next basin.

9
DR. LOVE'S LOST BONE BED

I hate leaving the Bighorn Basin, but it's never a boring experience. Since the basin is like a big walled-in depression, you either have to drive over a mountain range or through a canyon to get out of the place. Both of these experiences are fun for a person who likes rocks. The Wind River Canyon appears about five miles south of Thermopolis. Ray and I stopped at the mouth of the canyon, where the river issues from a giant gash in the otherwise smooth face of the Owl Creek Mountains. The rock formation at the canyon mouth is the Permian Phosphoria. This is the layer of rock that has produced *Helicoprion*, the whorl-toothed shark that had captured Ray's creative energies for over a decade. We stopped for a hero photo and then entered the canyon and headed uphill toward the Wind River Basin.

The Wind River Canyon is another one of those magical places where thousands of feet of layered rock are on public display. The stacked formations dip to the north into the Bighorn Basin, so as we drove south, we drove into lower and older layers. It's a disorienting place, since the river is flowing north and the rock layers are dipping to the north. The trick is that the gradient of the river is shallower than the dip of the rock layers, so even though southbound drivers are going uphill, it feels like they're going downhill. I had fun confusing Ray with this conundrum as

we drove uphill and back in time. Each time we crossed a contact between the formations, I'd call out the name of the layer and its age: Phosphoria-Permian, Tensleep-Pennsylvanian, Madison-Mississippian, Jefferson-Devonian, Three Forks-Devonian, Bighorn-Ordovician, Snowy Mountain-Cambrian, Pilgrim-Cambrian, Gallatin-Cambrian, Flathead-Cambrian, and finally the tortured 2.9-billion-year-old basement rock of the Owl Creek Mountains. Ray thought that my singsong recitation sounded sort of like poetry. It didn't, but we did drive right through the middle of nearly 3 billion years of rock in less than 30 minutes. There are not many places on the planet where you can do that.

Halfway through the canyon, we passed a giant cubic chunk of Bighorn Dolomite that had tumbled into the river. It looked like someone had tossed a huge die. "Bighorn Dolomite," Ray mused, "it sounds like a country band." Finally, we crossed over the great unconformity between the layers rocks and the much older black and twisted Precambrian basement rock and drove through a pair of tunnels and emerged into the blinding sun of the Wind River Basin. A dam at the south end of the basin was built in a zone of intense faults that formed as the Owl Creek Mountains were thrust up during the Laramide Orogeny, about 58 million years ago. The two-mile stretch of road is well

Pablo Puerta, the Patagonian fossil-finding machine, standing on a stack of crushed fossil turtles.

known as a war zone for summer geology field schools, and it's not unusual to see bands of confused students wandering along the side of the highway trying to make sense of the gigantic jumble of faulted blocks.

The July day had warmed considerably, and we decided to take a dip in the reservoir. We started to wade in but the water was disappointingly warm and full of floating nasty globs that Ray thought looked like sewage. I thought the smelly clumps were some sort of algae, but it felt like we were making a bad decision, so we aborted the dip. Feeling not at all refreshed, we headed for the junction town of Shoshoni.

I've always had a soft spot for Shoshoni because it used to have a good rock shop. It used to be that rock shops were common sights in small western towns. They were staffed by rock hounds who had quit their day jobs, converted a garage into a store, and lived the petrified version of the American dream. Once, passing through Shoshoni when I was 13, I met a kid at the gas station/rock shop who had found a perfect red jasper tomahawk head. He wanted a Stingray bike, and there was one available down the street for $12. After a strident negotiation, and with financial backing from my mom, the kid had his bike and I had my tomahawk head. I had begun to learn the art of the rock deal.

Anyone who travels Wyoming today knows that the Yellowstone Drugstore in Shoshoni is a mandatory stop. In this otherwise boarded-up Wyoming desert town, the drugstore is an ice-creamy oasis. Always packed with Yellowstone-bound tourists, the place appears to be staffed with every smiley-faced teenager in the county. A chalkboard on the wall attempts to tally the insane amount of ice cream that's churned into glacially thick, headache-inducing milkshakes that cannot be sucked through a straw. Their biggest year was 2000, when this hardworking band of rural teenagers served up 65,590 milkshakes and malts (we're talking about more than 13,000 gallons of ice cream!), with a record 727 on one day in late May. I always make it a point to contribute to this grand effort, and Ray was easy to convince.

We spooned our malts and pondered the fork in the road. If we headed east toward Casper, we would pass into the flat center of the Wind River Basin. For me, this was a place full of memories of fine fossil digs and Wyoming lore. Just east of Shoshoni is a string of tiny 10-person towns: Moneta, Arminto, Hiland, Lost Cabin, and Lysite. It was here in the 1980s that Richard Stucky, my boss at the Denver Museum, had made his career-capping discovery of a limestone lens full of the skulls and skeletons of tiny bush-babylike Eocene primates. In 1993, I'd followed him there to seek the fossils of the forest that had hosted these arboreal primates. Not far from Richard's desert camp, I found a splendid fossil rain forest deposit in the middle of the desolate plains. The Denver Museum had several glorious field seasons there with a site along the top of a small ridge. It yielded hundreds of exquisitely preserved leaves that clearly showed that this desert used to be a tropical rain forest.

During that time, we were building an exhibit about the evolution of life on Earth at the museum, and we took a lot of artists into the field to experience the joy of finding fossils. It was on one of these trips that artist John Gurche spotted a tiny piece of bone sticking out of a badland butte. As he dug in, the bone grew, and soon he was looking at the back side of a crocodile skull. I tried to speed the excavation process by chopping around the skull with my pickaxe, but was called off after I sent a crocodile vertebra flying over my shoulder. The gentler dental tools and brushes prevailed, and the complete skull, jaws, and partial skeleton emerged from the butte. When we got the specimen back to the museum and cleaned it, we realized that we had collected the first good *Diplocynodon* skull from North America. This is a crocodile best known as living during the Eocene in what is now France. Its presence in Wyoming was evidence that in the warm Eocene world, animals could wander from Shoshoni to Paris across a warm and forested land bridge that is now Greenland, Iceland, and Scandinavia.

Pablo the Patagonian had joined us for these digs. He had prevailed on the local ranchers to provide us with a pair of lambs that he slaughtered after murmuring, "Sorry, baby, but I have to eat." He roasted both lambs on iron crosses over an open fire. Throughout the whole process, he sang and sprinkled pepper water over the

splayed carcasses to keep them moist and tasty. The lambs took hours to cook, and our large museum crowd grew rowdier and rowdier. No harm, we thought; we are, after all, more than 50 miles from the middle of nowhere. Just then, the local sheriff pulled into camp in his dented Blazer. We were incredulous that our remote party had engendered complaints, but it turned out that the sheriff had heard tell of a Patagonian cooking lamb, and he wanted to taste some for himself. The party continued.

Since we had an appointment with Dr. Love, Ray and I turned west, following the Wind River, and aimed the truck for Dinwoody Canyon. David Love was born in 1913 on the Wind River Indian Reservation, and he always had a soft spot for the landscape formed by the intersection of the Wind River Mountains and the windblasted Wind River Basin. For most of the last half of his long life, Dave and his wife, Jane, maintained a summer place in Jackson Hole, but the increasing Aspenization of Jackson finally drove them east into Dinwoody Canyon. Our plan was to stop and chat with Dave and Jane and then make a dash over Togwotee Pass and into Jackson Hole.

We got to the Loves' cabin in the middle of the afternoon, but nobody was around, so we grabbed some pops from our cooler and made ourselves at home on the porch. I started to tell Ray my stories about Dave Love.

My relationship with Dave had grown since I marched into his office in 1991 and demanded my uncle's dinosaur tooth back. I'd taken to driving the two hours from Denver to Laramie every few months to sit in his large and jumbled office and talk about Wyoming geology. He was happy to see me because he loved to talk about his life as a geologist in Wyoming. Dave amazed me with his encyclopedic knowledge of the state. I was sitting at the feet of the master, trying to learn what he knew so that I could apply it to my study of fossil plants. Aware of this, he indulged me for hours.

Dave had a lazy Susan that held his notebooks from 60 years of fieldwork. I'd prompt him, asking if he ever saw fossil leaves in Cretaceous rocks in a certain part of the state, and then I would take notes furiously as he spooled off stories, occasionally stopping to spin the lazy Susan and pull out the appropriate notebook, from, say, 1941, and flip it open to the right page. The notebooks referenced slides, and he would pull these too, showing me the precise images I hoped to see. It was like having a conversation with the Earth. I felt immensely privileged.

Dave told me about an amazing fossil leaf site from the Late Cretaceous Harebell Formation in the Bridger-Teton Wilderness, where he had collected in the early 1940s. He'd taken Olaus and Mardy Murie, the famous wolf biologists, there, and they had filled a barrel full of fossils and shipped them back to the Smithsonian. Later, I visited the Smithsonian and pored over the accession records. No barrel had arrived in that year. I checked back with Dave, and he rechecked his notes. They were correct, and we surmised that the barrel had been lost in shipping. There was no reason why I couldn't revisit the site and collect my own barrelful. So in 1991, when my friend Jason Hicks and I were passing through Jackson, we decided to follow Dave's notes and see if we could relocate his fossil leaf site.

The notes led us on a long hike along a raging creek. After three miles, we came to the mouth of a canyon and had to ford the creek issuing out it. As I stood on the rocky bank looking for a place to cross, I stepped on a round branch that rolled under my foot and sat me down hard on the ground. I reached out to break my fall, but my open palm smacked onto the branch that felled me. There was a sharp pain in my palm, and I realized that I had impaled it on a broken, spiky, pencil-thick twig that projected up from the branch. Carefully, I pulled my hand off the branch and looked into my palm. I could see anatomy, including some strange white stuff, and I was instantly queasy. Fortunately, there wasn't too much blood. We didn't have a first aid kit so Hicks helped me wrap a bandanna around the wound.

We forded the creek and, following Dave's instructions, climbed up a steep ridge through a recently burnt lodgepole pine forest. The soil was thin, and the underlying bedrock was the conglomerate of the Harebell Formation, which is composed of potato-sized round cobbles. It was like climbing a hill of big ball bearings, and I fell several more times. After a few miles struggling up through this steep, surreal, and charred landscape, my hand was throbbing and my spirits were sagging.

Saurexallopus lovei, a four-toed Cretaceous dinosaur known only from its tracks.

I couldn't make sense of the notes I had taken during my conversation with Dave, and the sweet concept of a barrelful of fossils began to fade.

About that time, we spotted a small pile of angular sandstone blocks at the bottom of a steep 20-foot embankment. Hicks scampered down the wall like a squirrel. I took my time and about halfway down my foot rolled on a round cobble, and I rode the rest of the way to the bottom on the seat of my pants. Instinctively, I had extended my hands to slow my descent. When I stopped moving, I realized that I had packed dirt and gravel into the hole in my wounded palm. While I was stomping around and cursing my bad luck, Hicks went over to look at the blocks. Excitedly, he called out, "Check out these fossil worm tracks!" I went to where he knelt over a boulder and looked at the block, which did indeed have fossil worm tracks. But what Hicks hadn't noticed was that the worm tracks went up and over a spectacular set of dinosaur tracks. I forgot my hand and fell to my knees.

The foot-long tracks had four toes, each tipped with a sharp-clawed impression. Clearly these were dinosaur tracks, but they were also clearly from an unknown species. The Harebell Formation is from the last 5 million years of the Cretaceous, and I wasn't aware of any dinosaur from that time that had four toes. We collected two of the best specimens and hauled them back down the trail, across the creek, and into Jackson, where our friends were having a large party. We arrived about ten o'clock at night, dirty and bloody, but carrying a splendid fossil. The party stopped when we walked into the center of the room and set the briefcase-sized rock on the coffee table. Jackson Hole had its first dinosaur.

We later named this dinosaur in honor of Dave Love. The name, *Saurexallopus lovei*, means "Love's strange reptile foot." Dave, while flattered, was also annoyed that he had missed the tracks when he was in the area in 1941. So it was that in his early 80s, he too climbed the ball-bearing hill to see the site of the four-toed tracks.

Just as I finished telling this tale to Ray, Dave and Jane rolled up their driveway in a green Subaru. They were delighted to see us, and Jane soon replaced our

warm pops with ice-cold beers and crackers. Right off the bat, they told us that one of Dave's neighbors had found a fine fossil footprint in the red Triassic rock at the mouth of Dinwoody Canyon. This was interesting news, as these rocks were notoriously stingy with their fossils. I remember reading a scientific paper from the 1930s about them that was more of a lament than a thesis. The authors talked about how they had searched the Chugwater and Popo Agie (pronounced "poh-pah-jah") formations around Riverton for months, expecting to find crocodile-like phytosaurs, small-headed herbivorous armored aetosaurs, and giant-salamander-like metoposaurs, the kind of fossils that are relatively common in rocks of the same age in New Mexico and Arizona. Instead, they found next to nothing. They talked about walking endless ridges for weeks, systematically dividing the outcrop and searching with great care. They knew that they wouldn't find anything if they didn't look, but they looked hard and still didn't find anything.

So Dave's neighbor was especially lucky. Dave showed us a picture of the amazing five-toed track. It was about the size of my hand. Like the tracks that Hicks and I had found, this track was another example of a creature

Born in the Wind River Basin, Dave Love became Wyoming's most famous geologist.

that left only footprints. A mystery like so many fossils tracks, this was just one more example of how little we know about the past and how many more fossils there are to find and understand.

Love had previously shown me tantalizing images of a dinosaur bone bed from the high country south of Yellowstone. The images were faded but the jumble of large bones was unmistakable. I asked him if a paleontologist had ever visited the site, and he was sure that none had. I had told this story to Ray, and he started obsessing about Dr. Love's Lost Bone Bed. Ray quizzed Dave about the lost bone bed that afternoon. But Dave, in an impish way, derailed the topic and started talking about the evolution of the jackalope. He had a particularly fine mounted specimen of this uniquely western beast on his wall. It was clear that the conversation was going to be light. And so it was. We said our farewells and headed west. This was the last time I ever saw Dave. He died the next year.

I was quarrying fossils in Utah the week Dave died, and I drove alone across the full width of Wyoming to attend his memorial service in Laramie. It was amazing to me how many outcrops I passed on that drive that he and I had discussed. I often think of Dave: his encyclopedic knowledge of Wyoming, his irreverent whimsy, and all the stories he didn't tell me. The big empty state of Wyoming will be a whole lot emptier without him. I've lost Dave, but one day I'm going to find Love's Lost Bone Bed.

West of Dinwoody, the valley of the Wind River narrows and becomes wooded. The evening sun lit up the brilliant red cliffs of the Wind River Formation on the north side of the road, and we rolled into Dubois just as the evening light was going flat. We stopped for gas, and I looked across the street at the old wooden sidewalk and storefronts that are typical of old western towns. Even though it was nearly nine, I noticed that there was a light on in the Two Oceans Bookstore.

What seems an odd name for a bookstore in the mountains of Wyoming is not so wrong when you realize that Togwotee Pass between Dubois and Jackson follows the Continental Divide. The water in the Wind River

eventually makes it to New Orleans, while the water in the Snake River, which flows through Jackson Hole, will eventually reach the Pacific at Astoria, Oregon. I remembered that the owner of the bookstore had reprinted one of Dave Love's classic monographs on the geology of the Absaroka Mountains, so I wandered across the street and knocked on the locked glass door.

After a few moments, Anna Moschiki answered and invited us in. The store was closed, but she was picking something up and was curious about us. Ray was flattered to find that the store carried copies of his books, and Anna put him to work signing them. One thing led to another, and pretty soon we were headed to Anna's house up the valley for a beer and the chance to meet her husband, Mike Kinney. Mike is straight from central casting, a quintessential cowpoke with a big ol' mustache, and their cabin on the banks of the Wind River is packed with western literature. Soon the beer turned into a fine meal of bangers and mash, and they wouldn't hear of us pushing on. We spent the night talking about fossils and stories that wolves from Yellowstone had been killing local dogs and horses. Anna and Mike, friends of the Loves, had been to visit the four-toed track site in the Bridger-Teton Wilderness and had found several more prints themselves. We talked on into the night about the possibility of someday mounting a packhorse expedition to find Dr. Love's Lost Bone Bed. The next morning, we had some strong mountain-cowboy coffee, said farewell to our new friends, and hit the road.

The valley of the upper Wind River is densely forested and densely fossiliferous. Several Eocene formations crop out amongst the trees. These outcrops contain fossil leaves and mammals from a time when tropical rain forests covered the slopes of the Wind River Mountains. This area has attracted generations of paleobotanists. One of my favorites is Roland Brown from the U.S. Geological Survey. Brown was a notorious penny pincher who wore the same pair of pants for many years, gradually adding panels of cloth as he added pounds of flesh. Like many a paleobotanist, he was a deft hand with a trimming hammer. When you collect hundreds of fossil leaves, you want to minimize the rock-to-fossil ratio to lighten your load. On an expedition in the Wind River Valley, Brown was collecting fossil leaves when he found a splendid fossil bird feather on a big slab of rock. Much to the horror of the people he was with, he proceeded to trim the rock until it was only slightly larger than the feather. Fossil trimming is my 10,000-hour skill and I've broken many a fossil along the way. That is why glue is always part of the paleobotany tool kit and why my social media handle is @leafdoctor.

We crested Togwotee Pass and coasted down into Jackson Hole. The outcrops on the west side of the pass are from the Harebell Formation and have yielded the teeth of the dinosaur *Leptoceratops,* a modest little cousin

of *Triceratops* that seems to be more common in deposits formed near ancient mountain ranges. Maybe this little guy was the Rocky Mountain sheep of his day.

Yellowstone is a volcanic wonderland that started erupting more than 50 million years ago and has been catastrophically active as recently as 639,000 years ago. An army expedition into Yellowstone in 1871 confirmed rumors of hot springs and geysers, which led to the place being called "Colter's Hell" after one particularly loquacious mountain man. In 1871, 42-year-old Ferdinand Vandiveer Hayden led a group that included scientists, surveyors, the painter Thomas Moran, and the photographer William Henry Jackson into Yellowstone. Hayden had been exploring the West since he graduated from college,

Bluffs of the Eocene Wind River Formation near Dubois, Wyoming.

and he already had a string of discoveries under his belt. On the banks of the Missouri River on one of his first trips west in 1854, he collected teeth of the first dinosaur ever found in North America. He loved natural history, and even though his primary goal was to survey land and resources, he sensed that the general public really wanted to hear about and see images of the amazing American West. On this trip, Hayden had some great artists with him and he made the most of their images. Armed with paintings and photographs, he convinced Congress to name Yellowstone as the nation's first national park within months of his expedition. Voted into existence on March 1, 1872, Yellowstone started the national park movement.

In the northeast corner of what would become the park, Hayden's men discovered a series of long ridges that were studded with petrified trees. Up to eight feet in diameter, the agatized trunks stand as high as 12 feet. But they aren't an isolated find: layer after layer of buried forests lie beneath them. There are at least 27, and perhaps as many as 60 stacked forests along Specimen and Amethyst ridges. During the formation of the Absaroka Mountains, volcanism and mudslides alternated in a catastrophic continuum. Subtropical broadleaf forests grew on the flanks of the volcanoes, only to be killed and buried during eruptions. One of the world's greatest fossil sites, these petrified forests have languished in obscurity, overshadowed by the sexier sights of Old Faithful and grizzly bears. The park management has not made a
point of pointing them out, and most park visitors who drive through the lovely Lamar Valley have no idea what they're missing.

In 1960, Erling Dorf published a popular description of the stacked forests and began to restudy the publications produced by the Hayden surveys. Erling's diagram of 27 stacked fossil forests caused a lot of consternation to creationists, who pegged the age of the Earth at around 6,000 years. If you apply simple math to 27 stacked forests, each several hundred years old based on growth rings, it's pretty easy to use up all of Earth's history in a single hill in Wyoming.

Beginning in the early 1970s, and in clear response

to Dorf's findings, Yellowstone saw a backlash research effort led by Seventh Day Adventist creation scientists who were intent on finding a mechanism to bury and stack 27 forests in less time that it took to grow 27 sequential forests. The idea they came up with was that the forests slid into place like a stack of coasters and that all of the layers came from a single forest that was disrupted by huge earthquakes and mudslides. Growing up in a Seventh Day Adventist family, I first read about these petrified forests in church bulletins. I remember pained discussions in Sabbath school about the implications of these stacked forests and their conflict with Bishop James Ussher's 1658 proclamation that the first day of Creation was Sunday, October 23, 4004 BCE.

Interest in the Yellowstone fossils led the Seventh Day Adventist researchers to Mount Saint Helens after the great eruption of 1980. Here they made an interesting discovery. The blast from the eruption had blown down thousands of giant trees, and many of these trees ended up floating on the surface of nearby Spirit Lake. Eventually, the trees got waterlogged and started to sink. Since the base of the trunks were wider and heavier, they sank first, and many of the trees ended up floating vertically in the water column. Some of the trees sank and embedded themselves in the muddy lake bottom, creating a submerged standing forest that was buried where it didn't grow. In creationists' desire to compress time and confirm a biblical catastrophe, they discovered an interesting volcanic process. Despite this discovery, the presence of fossil soils and the fact that many of Yellowstone's petrified trees were actually rooted in the fossil soils showed that the Yellowstone forests really were stacked sequential forests, not transported ones. On a 4.567-billion-year-old planet, 6,000 years is simply not enough time to get serious work done.

It was ironic that my fundamentalist upbringing brought me into contact with the study of strata long before most kids even know that rock layers were once landscapes. In a church where Noah's flood was the default answer for any geologic problem, I started to think for myself. I remember standing by a road in British

THE AMAZING STACKED FOSSIL FOREST OF YELLOWSTONE NATIONAL PARK

Columbia collecting fossil fish from a hill of paper shale. It occurred to me that each thin layer, or varve, might have been a year's deposit. I counted the varves in a foot of rock and guessed the thickness of the road cut. By my own teenage calculation, I was looking at 30,000 years. All teenagers rebel against something, and that day I sank Noah's ark. To their credit, my naturalist parents supported my observations and didn't let the dogma of their upbringing affect the progress of mine.

Ray and I turned left at Moran Junction and drove the road through Jackson Hole past the exquisite Teton Mountain Range toward Jackson. This is one of the prettiest places in North America. My mom used to bring me to Jackson Hole in the summer, and it was here, in a parking lot that I learned the pure joy of successful searching. I must have been about 10 years old at the time, and I got it in my mind that I could find an arrowhead. My mom was patient, but she had limits, so she told me I had five minutes to find an arrowhead and then we had to leave. I sprinted out of the car and began scanning the ground. Somehow I knew that I would have better luck in the bushes than on the open ground. I crawled under a dense willow thicket and there, to my amazement, lay a beautiful obsidian arrowhead. I rushed back, well within the five-minute limit. My mom could not have been more surprised. Ever since that time, I have set out to find things with the faith that I will find them. For a paleontologist, this is a useful mind-set.

I can also remember riding the gondola at Teton Village to the top of the range and finding fossil marine brachiopods more than 10,000 feet above sea level. This was startling to me at the time, but now I know that mountains and seas come and go and marine fossils on mountaintops are the norm rather than an exception.

Jackson is a western tourist town that has always had a good rock shop. It also has a town square with famous archways of elk antlers and a staged gunfight that breaks out every summer afternoon. The beauty of the surrounding landscape has long drawn people to buy property and second homes. In 2000, the median house price in Teton County was more than $565,000, a tenfold increase from 1980. By 2022, the median sales price hit $4 million. The result is yet another western town where the townspeople can't afford to live. We could easily see why Dave and Jane Love had moved east to Dinwoody.

We left Jackson around five and headed over Teton Pass to Victor, Idaho. We had arranged to meet my friends Susan and Mayo at their house in the country. Susan greeted us with a freshly baked apple pie that was topped with a toothy dough dinosaur. Mayo's murdering ways supplied us with a dinner of dinosaur descendants in the form of a brace of tasty grouse. After a splendid meal, I hit the pillow hard. But Ray was restless, because the next day we were headed to a place that he had longed to visit for years.

There is a bizarre fossil known as *Helicoprion* that had possessed Ray's artistic soul since he first encountered it in the late 1980s. It's a whorled thing, ranging from a few inches to a few feet in diameter. From even a short distance, *Helicoprion* looks like a beautifully coiled ammonite. But let your eyes focus, and you'll see that the whorl is made of teeth. This wicked spiral is the business end of a Permian shark, and some of the best specimens in the world come from the eastern edge of Idaho. Ray first learned of *Helicoprion* when he saw one being used as a doorstop at the Natural History Museum of Los Angeles County. Ray can be obsessive, and *Helicoprion* became his obsession. He'd visited whorl-tooth specialists and worked with them to make credible reconstructions of the animals, but he'd never found one in the wild.

Susan and Mayo joined us the next day, and we drove south along a series of long valleys. The Idaho-Wyoming border country is composed of long linear ridges that run north and south and are the result of a messy geologic bust-up where nearly flat faults sliced and shifted rock layers like a deck of cards. This is the overthrust belt, a place where the layers of sedimentary rock are faulted but the basement rock doesn't come to the surface. Geologists call this kind of deformation thin-skinned, as opposed to the thick-skinned Rocky Mountains where the basement rock is shoved to the surface. Some of the same

A whorl of
shark teeth.

he'd seen his share of *Helicoprion*. Ray was nervous and excited as we closed in on his "whorl-tooth blind date." We followed the instructions that Jay had given over the phone and soon pulled up in front of a two-story house at the edge of a foggy cow pasture. Susan, Mayo, and I hung back as Ray met Jay. Jay, a lean and lanky 60-year-old with a trimmed gray beard, caught his first glimpse of Ray and said, "I knew that you would have a gray beard just like mine." He invited us in, introduced us to his wife, and showed us some of his rock collection. Out in his yard, he had a giant concretion from the mine. Apparently, the whorls are found in the middle of concretions, just like the ammonites Ray and I had collected at Rock River. But Jay's was a gigantic concretion, the size of a beanbag chair. And it was very, very hard. I started calculating the size of the sledgehammer that would be needed to crack it open and soon realized that I wouldn't be able to lift such a hammer.

Jay took us to a local café for a hearty lunch. Ray ordered a chicken-fried steak and burrowed deeply into conversation with Jay. The date was going very well indeed, and Ray was bonding with his whorl-tooth brother. We discovered that in his 38 years in the mine, Jay had seen only six *Helicoprion* whorls, but we were feeling lucky, so we decided to visit the mine after lunch.

There was excitement in the air as we drove back into Idaho and up the road to the Smoky Canyon Mine. As we approached, we saw huge piles of concretions lining the road. These giant indestructible Milk Duds were a pain to the miners, but they made good riprap, so the mine used them to stabilize slopes. We stopped and looked at a few, but soon realized that there was no way we would ever get them open. Ray looked like he might burst into tears as it dawned on him that there must be shark spirals inside some of the hundreds of impregnable beanbag boulders that littered the hill, but he was never going to see them.

We parked at the entrance to the mine and signed in at the office where a big bear of a man gave us a safety

layers are here, but instead of being laid out in an orderly fashion, they're sliced like deli meat, and it can be real tough to know where you are in the geologic column. This, of course, made me edgy and unhappy. Ray didn't really care about the sordid details of the rock formations; he was fixated on his freak show of a shark.

Ray had a *Helicoprion* pen pal he'd been corresponding with for years, a fellow by the name of Jay Muir who worked in a phosphate mine near Afton, Wyoming. So we pulled into the home of the world's largest elk antler arch around noon, planning to meet up with him. The Phosphoria Formation, not surprisingly, has a lot of phosphate, the legacy of its origin as mud on the bottom of a Permian sea. Jay had worked the mines for 38 years, and test and issued us bright pink hard hats and shockingly yellow steel toes for our shoes. He was the local wrestling coach, and he told us to watch out for his boy, Rulon, in the upcoming Olympics. You might remember that Rulon Gardner went on to win the gold medal in 2000 by unexpectedly upending a giant Russian wrestler. At the time, we didn't think anything about it.

Garbed in our absurd pink hats, we toured the garage where the giant mining trucks were repaired. Ray was wondering if the ridiculously colored safety gear wasn't a way to humiliate city slicker visitors. We thought we saw more than one dust-covered miner chuckling to himself.

The immense mine produced phosphate ore, which was ground into a slurry and pumped to Pocatello, where it's used as fertilizer. It's sort of odd to think that your French fries are made from potatoes that are fertilized by ground-up Permian whorl-toothed sharks. Piles of concretions were all over the place. Some of them were eight feet in diameter. We scrambled over the mine spoils, convinced that we would find a whorl-tooth in a naturally cracked concretion. We hunted until it was starting to get dark

and found a few fragments of fossil shell, but the sharks were elusive and our hammers were simply not up to the task. Reluctantly and with damp spirits, we left the mine and headed back to Afton without having seen even a fragment of a whorl-tooth. It was then that Jay mentioned that, years before, he had given one to a local elementary school.

The next morning, Jay, Ray, and I drove to the Osmond Elementary School, named after Donny and Marie's grandfather. School was in session, but we soon found the principal, Kelly Tolman, and asked him where he kept his shark. He really didn't know what we were talking about but said that there was a big rock out by the Dumpster. With mounting excitement, we rounded the building, and there, lying forlornly on the asphalt next to the Dumpster, was half of a big Phosphoria concretion. On its surface was a glorious 18-inch whorl. Ray fell to his knees. Kids had been jumping on the rock for years and there was some blue paint splattered on it, but nothing could hide the fact that this was still a fine fossil. It even held the impression of the shark's skin. Ray explained the significance of the fossil to Principal Tolman, who

The children of Osmond Elementary doing the sign of the whorl.

soon had us in a classroom in front of all of the teachers. Then the kids were released from class for an impromptu whorl-tooth presentation, and Ray got a stick of chalk and starting drawing a full-scale whorl-tooth on the concrete playground. Excitement was building as we explained to them just how rare these fossils were. By this time, Ray and Jay had fully bonded and were sharing secret whorl-tooth handshakes that looked like a cross between gang signs and rotating dishwashing maneuvers.

The team name for Osmond Elementary is the Star Valley Braves. Apparently it used to be Star Valley Cheesemakers, but that was just asking for a beating. Ray started berating the poor elementary school principal, telling him that if they had changed the name once, they could change it again, this time to the Osmond Elementary Fighting Whorl-Tooths. He promised to design a hip logo that could grace school jerseys. Principal Tolman was diplomatic, and we left town thinking we might actually have a chance at changing the mascot.

Over the next several years, Ray went on a real whorl-tooth bender. He drew dozens of images of the swirly demons and began to recruit scientists to join him in his obsession. He finally snared Idaho State University paleontologist Leif Tapanila, who is known for his work on fossil clams. Together they rigged up a metal model of a whorl-shark head and used it to chop whole salmon

and watermelons in half on YouTube. Eventually he built a traveling whorl-tooth shark exhibit. The marketing team got to him and the next thing I knew, Ray had rebranded the whorl-tooth shark as the buzz-saw shark. I guess it's a good think that Principal Tolman didn't go for the new mascot.

Ray and his whorl-tooth buddy, Jay Muir.

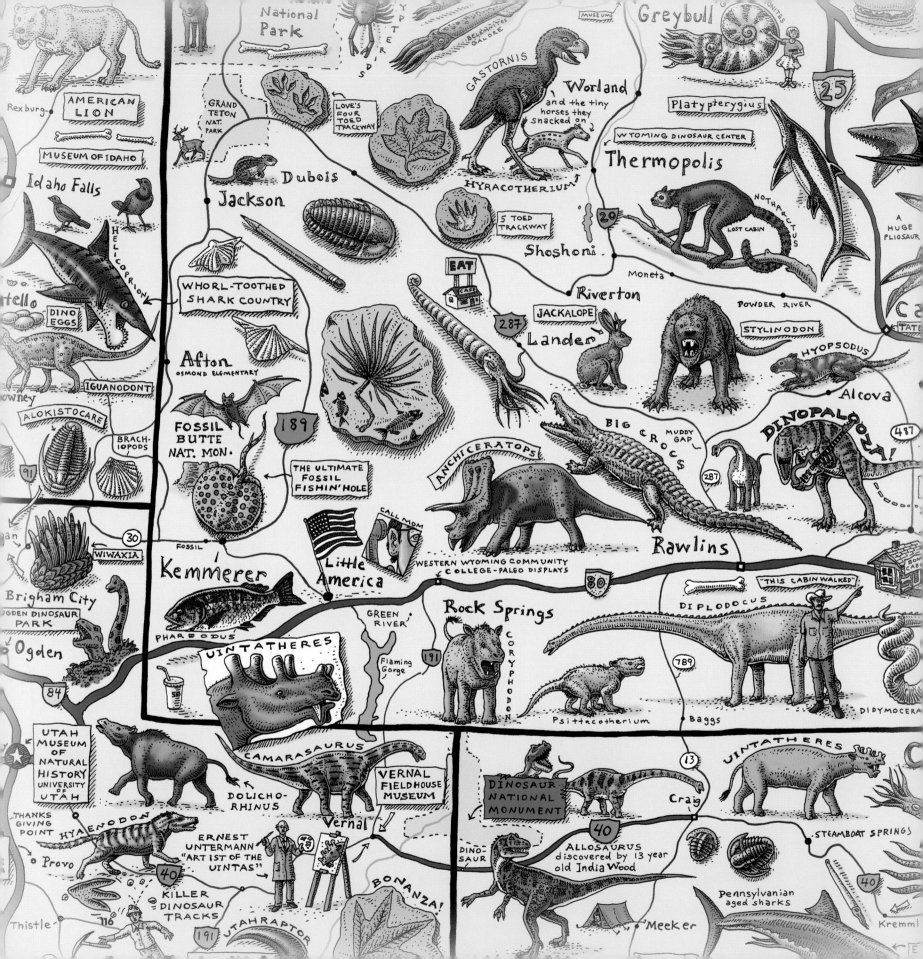

10
SURF AND TURF

There are few fossil sites more famous than the Green River fish beds near Kemmerer, Wyoming. The Green River Formation is a huge sequence of lake bed sediments that started to form as the Rocky Mountains were uplifted during the Paleocene. The mountains at the corner of Wyoming, Utah, and Colorado formed in such a way as to surround a huge area of low-lying land between them. It was a time when the climate was warm and wet and the runoff from the mountains had nowhere to go but into the land between the mountains. The great rivers that would eventually carve the Continental Divide didn't yet affect this area, and a series of huge lakes formed as the space between the mountains filled with water.

The southwest corner of Wyoming is also where the modern Rockies are their most gentle and aren't a formidable barrier to travelers. As a result, this area saw some of the first westbound explorers and, by the 1840s, became the path for the Oregon Trail. For this reason, Green River fossils were discovered very early in the game. The first paleontologists in southwestern Wyoming came by rail, and they couldn't miss the spectacular cliffs that paralleled the rail line (now Interstate 80). When they found fish in many of the layers, they realized that they were looking at the bottom of a huge lake.

Kemmerer, a hamlet that advertises itself as a "small town with a population of 103 pronghorn antelope, 113,000 fossil fish, and 3,000 people," is close to the Idaho line and a 50-million-year-old lake that formed between two ridges of the overthrust belt. Somehow, the lake's chemistry and sediment input were just right for fish to live, die, and be buried, three necessary ingredients for any fossil. A zone in the middle of the formation known

as the 18-inch layer is the main pay zone. The fish from this layer are spectacular fossils, chocolaty brown bone on a creamy matrix that pulls up in sturdy table-sized sheets. The prepared fossils look like works of art, and they're often sold in frames.

There's skill and persistence in the excavation and preparation of these spectacular fossils, because the shale doesn't split cleanly away from the bone. Usually a few millimeters of the matrix coat the bone in such a way that you see a ghostly raised apparition of a fish skeleton in the rock. It takes a sure hand with a pocketknife, needle, or stiff brush and countless hours of patience to flake away the matrix and expose the fish.

Digging into the 18-inch layer is tricky. The rock is nearly white, so digging during the day can cause you to miss the subtle fossils due to the blinding reflection of the Wyoming sun. Some of the old-timers will get around this by working at night, using the low-angle light of a lantern to highlight the obscure lime-draped skeletons. One of my

FISHES OF THE GREEN RIVER FORMATION

1. *Amphiplaga* (Trout Perch)
2. *Diplomystus*
3. *Knightia*
4. *Gosiutichthys*
5. *Priscacara*
6. *Eohiodon* (Mooneye)
7. *Mioplosus*
8. *Lepisosteus* (Gar)
9. *Crossopholis* (Paddlefish)
10. *Notogoneus*
11. *Phareodus*
12. *Amia* (Bowfin)
13. *Heliobatis* (Stingray)
14. *Asterotrygon* (Stingray)

most surreal nights was spent digging the 18-inch layer from 10 P.M. until 4 A.M. with 82-year-old Lloyd Gunther and his 80-year-old wife, Freida. I was 31 at the time and happy to be there, but also dead tired and bone cold. It was all I could do to keep up with the tenacious octogenarians as they popped fish after fish out of the old lake bed and gave me a good look at true marital bliss.

There's another zone, more suited for lazy diggers, called the "split fish" layer. Here the bone is apparent when the rock splits, but the fossils are not as pristine or beautiful as those from the 18-inch layer. Digging the split fish layer is like taking candy from a baby, and it's easy to end up with a school of fossil fish.

The fossil quarries in the valley have different legal statuses, and the two most common are those on private land and those on Wyoming state land. No regulations apply to the former, but the latter are administered by the Wyoming state geologist. Fossils deemed rare are supposed to be turned over to the state collections in Laramie. The federal government also manages BLM land in the valley, and, in 1972 Fossil Butte National Monument was commissioned from 8,720 acres of federal land. The stories about the fish diggers and their disputes are enough to fill a book, and Ray and I talked about all of this as we drove into the basin from Idaho. We'd both been here before, but we knew different diggers.

Our first stop was Carl Ulrich's house and shop at

Stingrays of Wyoming.

the head of the road that leads to the monument. Carl has been prying up shale and selling fish since 1947, and he's the senior fish digger in the valley. He has a prime quarry lease on state land and a skilled hand for preparing the fossils. He prepares the fish with an eye for artistic composition and then frames them like fine paintings. You can always tell an Ulrich fossil because Carl signs the lower right-hand corner, just like an artist signs a piece of art. Carl also had the bright idea of selling unprepared fish to would-be junior paleontologists. I spent much of my youth needling away on Carl's U-prep-it fish. I can tell you firsthand that Carl is one very patient person.

I met Carl's son, Wally, in Jackson when I was 12. He treated me like an adult, taking me to lunch and talking fossils at a local deli. Thirty-two years later, I walked into his Jackson fossil gallery with my new wife. He glanced at me for hardly a second before saying, "You've grown since I saw you last." Apparently, he kept track of the kids he took to lunch. Wally is also in the fish business, but he has branched into selling fossil fish from the tougher brown layers of shale, where the fish are hard to extract. He treats the stone like wood, sawing and planing it into slabs, exposing the fish by sanding and polishing the stone. The resulting pieces are hung as art or built into heavy but appealing furniture. The top of the bar at the Phantom Canyon Brewing Company in Colorado Springs is made of this stone, but few of the patrons realize that they are raising their pints from a surface that used to be the bottom of an ancient lake in Wyoming.

Many other families of diggers have also worked the valley, and more move in all the time, pulled by the lure of fossil fishing and the hope of finding something fantastically rare. Guys such as Jimmy Tynsky, Tom Lindgren, and Rick Hebden compete with the Ulrichs. They all dig for a living, but they also dig because they love finding the really rare stuff.

The meat and potatoes of the Green River fish business are five common species: *Knightia*, *Diplomystus*, *Priscacara*, *Phareodus*, and *Notogoneous*. They can be sold from the state leases without accounting. Species such as the big scaly garfish *Lepisosteus*, stingrays, paddlefish, and catfish are considerably rarer and are managed by the state of Wyoming. Of course, if you own or lease a quarry on private land, there are no restrictions on the rare fossils.

Besides the ubiquitous fish, there are some stunningly rare and fantastic plant and animal fossils from the 18-inch layer. These include whole alligators and crocodiles, turtles, complete palm fronds, birds, mammals, lizards, crawfish, insects, leaves, flowers, and even snakes. In 2004, Jimmy Tynsky, a longtime Green River digger, flipped over a four-foot slab and found a nearly perfect and wholly complete fossil horse skeleton. The little four-toed steed was laid out on the slab as though it were sleeping, and the rest of the slab was sprinkled with fossil fish in a sweet little surf-and-turf composition.

PERHAPS THE ULTIMATE SURF AND TURF FOSSIL

119

A lot of people think that this is one of the finest fossils ever found. When we wrote the first edition of this book, this fine fossil was tucked away in a safe spot somewhere in Kemmerer. In 2017, Jimmy Tynsky decided that he was ready to sell his prized horse and I was ready to buy it. It is now on permanent display in the David H. Koch Hall of Fossils-Deep Time at the Smithsonian's National Museum of Natural History where more than 5 million people a year have the chance to see it.

The market has always lusted for these rare beauties, and many of the very best fossils have been spirited out of the basin and into private collections. Sometimes they're photographed or cast, but other times they simply vanish from the public eye. Today, a rare Green River fossil can easily fetch a six-figure sale price, and some are even in the seven-figure range.

One of the best ever found was a fossil bat collected in 1935 by Clarence Cushman and given to Princeton in 1941. It's an exquisite gem of a fossil, only about three by six inches but utterly complete. Princeton paleontologist Glenn Jepsen took his time and prepared the bat from both sides, eventually describing details as fine as the hammer, anvil, and stirrup bones of the tiny mammal's inner ear in a paper in *Science* in 1966. Wally told me that Jepsen used to carry these tiny bones around in a vial. When Princeton University decided to toss in the towel on paleontology in 1984, they gave their fossil mammal collection to the Peabody Museum at Yale. Today, the bat resides in a secure safe in New Haven. A few years ago, I made an appointment with dinosaur paleontologist and Yale curator John Ostrom to view the fossil. After much fussing, he pulled the tiny slab from the safe and handed it to me. I was surprised how small and perfect it was. To my delight, there was a perfect little fossil flower right next to one of the bat's feet.

The fate of other Green River treasures is more obscure. A beautiful fossil boa constrictor was cast and replicated, but the whereabouts of the original are unknown. In the last decade alone, I've laid eyes on several bats, birds, and a couple of complete four-legged mammals. All of these specimens were destined for market and are now largely out of reach of science.

When Ray and I stopped to see Carl, we were amazed to discover that he had a small *Helicoprion* on display. Carl wasn't around, and the clerk had no idea where the fossil came from. Ray insinuated to the woman that he was prepared to pay any price for the whorl, but the hired help was well trained and politely rejected his entreaties. We admired a giant garfish that has graced Carl's shop for years and then headed for the national monument.

Fossil Butte National Monument has done an excellent job of working with both local commercial diggers and the scientists who do the primary research in the basin. Lance Grande, a bowfin fish specialist from the Field Museum in Chicago, is the reigning expert on the fossil fish of the Green River Formation, and his handbook to the fossils of the basin is like a bible to those who dig this pale rock. Lance has taken the time to get to know the diggers. As a result, the Field Museum has the best publicly owned Green River fossil collection.

We checked out the visitor's center, which does a good job of showing what Fossil Buttes looked like when it was at the bottom of a lake. I asked to see the collections, and we were told that the curator had gone and taken the keys with him. As we were about to give up and leave, I recognized a face I knew. Vince Santucci, who had been on the other side of the fence from Pete Larson in the Sue incident. Vince is an odd duck: part hard-nosed cop and part practicing paleontologist. I first met him when he worked at Petrified Forest National Monument in Arizona. There he set up stings to catch people who stole souvenir pieces of petrified wood. He

Fossils and the Man

As it presently stands, there are no laws that concern collecting fossils on private land. For this reason, there's a legitimate commercial industry of fossil collectors who work on land they own or lease. Different states have different laws and regulations for state-owned land, and each one is slightly different. In addition, there are several kinds of federal land—Bureau of Land Management, Forest Service, National Park, Bureau of Reclamation—and each of them has slightly different rules and regulations as well. Indian reservations also have their own rules.

Perhaps the most controversial type of federal land is managed by the Bureau of Land Management (BLM), whose mandate is multiple use. Presently, private individuals can collect reasonable amounts of plant and invertebrate fossils for personal use on BLM land, and scientists may collect significant invertebrate, plant, and vertebrate fossils, as long as they obtain a permit and give all of the fossils to a federal repository. No form of commercial paleontology is permitted.

The BLM was formed to manage the land that was left over after the more arable acreage was homesteaded and the forested land was placed under the jurisdiction of the U.S. Forest Service. BLM land comprises much of the fossiliferous badlands of the American West. Commercial diggers would like access to this land and argue for multiple use.

Amateur paleontologists are allowed to collect on BLM land but must beware that they don't accidentally collect vertebrate fossils. Museum and university paleontologists can collect on BLM land but must make sure their permits are in order. In 2009, President Barack Obama signed the Paleontological Resource Protection Act into law. This law still allows for casual collecting of common fossil plants and invertebrates on BLM land, but it is still a good idea to stop in at the local BLM office before you try it yourself.

oversaw experiments in which they would spray a section of ground with paint that was invisible to the naked eye but showed up well under special lights. Then they would rephotograph the section of ground over time and calculate how much wood had been removed by stealthy tourists. The results were staggering: tons of fossils were being hauled away each year. Vince had also been stationed at Big Badlands National Monument in South Dakota, where he began enforcing the unenforced laws against collecting fossils.

Ray and I spent a couple of hours listening to Vince talk about protection of fossils on federal land. One of the main fronts of the controversy over who should and should not access the fossils occurs in southwestern Wyoming. This is a landscape which is spectacularly barren, chock-full of vertebrate fossils, and largely managed by the BLM. Vince told us about a zealous flying sheriff who ran a sting operation called Operation Rockfish in which he flew over the southwest corner of Wyoming in a Cessna looking for people out illegally collecting fossil fish. The Green River Basin is a gigantic area where rock hounds have long collected the superabundant fossil herring known as *Knightia*. In one area near the tiny town of Farson, a fossil fish site looks like a heavily cratered World War I battlefield. But over the last 35 years, what was once common hobbyist practice has become illegal, and oblivious rock hounds have inadvertently become criminals. Crime and paleontology just seem like such odd bedfellows.

Ray and I talked about Rodney King and his infamous plea "Can't we all just get along?" It seemed silly to us that with so many fossils and so few people who love fossils, we can't find a better way to manage the resource. We both agreed that the best hope is clearly explaining the rules and regulations to as many people as possible.

We had lunch in Kemmerer and drove south along Fossil Ridge, a 50-mile-long ridge of the 92-million-year-old Frontier Sandstone. Fossil Ridge got its name because in places it is literally made of fossil oysters that lived along the shore of the Western Interior Sea. Many long ridges of Cretaceous sandstone in Wyoming are the remains of old beaches that have been tipped on end. At a place called Cumberland Gap, we cut through Fossil Ridge and into the Green River Basin. The Frontier Formation at Cumberland Gap is famous because the Frémont expedition of 1843 passed through this very spot. Here, Frémont himself found a layer full of spectacular fossil ferns.

We hoped to make it to Rock Springs by evening, so we sailed along, but we were passing over hallowed ground. The southern part of the Green River Basin is known as the Bridger Basin. The Bridger Basin was the site of one of the first mountain man rendezvous in 1825, and these crusty old beaver-trapping boys apparently knew a fine fossil when they saw one. Rumors of the region's fossils traveled back East. The completion of the Union Pacific Railroad in 1869 made it possible for an eastern academic to hop a train in Philadelphia and be in the Bridger Basin less than a week later. For this reason, the Bridger became a proving ground for the budding young field of American paleontology; nearly all of the important early workers cut their teeth on the badlands of the Bridger.

Joseph Leidy, naturalist extraordinaire from the Academy of Natural Sciences in Philadelphia, was one of the first, arriving in Bridger Basin in 1872. The blue-gray badlands of the Bridger are full of 50-million-year-old fossils, and here Leidy found the bizarre skulls of the knob-headed, saber-toothed herbivore known as the uintathere. Within a few

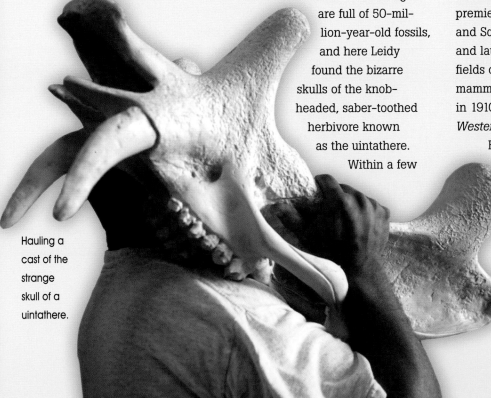

Hauling a cast of the strange skull of a uintathere.

years, both Edward Cope from Philadelphia and O. C. Marsh from Yale arrived in Fort Bridger and started collecting, describing, and naming uintatheres. In their haste to out-publish each other, they described fragments rather than complete skeletons and made a mess out of the scientific literature. Leidy grew disgusted with the antics of his junior colleagues and retired from the field, leaving Cope and Marsh to battle it out.

Cope's work resulted in the publication of a massive 1,009-page, 16-pound, 5-ounce volume, *The Vertebrata of the Tertiary Formations of the West*, published in 1883. This monster of a book has come to be known as Cope's Bible. Marsh, in order to outfox Cope, led a group of Yale students over the Uinta Mountains into the Uinta Basin in 1870, and he published the results of that trip in a two-volume set in 1886 that named even more uintathere species. In 1876, three Princeton undergraduates, avoiding exams by lounging in the shade along a canal in New Jersey, decided that rumors of fossils out West were just too good to be ignored and organized their own expedition. Two of these guys, Henry Fairfield Osborn and William Berryman Scott, were to become towering figures in the study of fossil mammals. Osborn went on to establish the American Museum of Natural History as the premier vertebrate paleontology museum in the world, and Scott returned to Princeton to build their program and launch a series of successful expeditions to the fossil fields of Patagonia. Each wrote a major book about fossil mammals: Osborn's *The Age of Mammals* was published in 1910 and Scott's *A History of Land Mammals in the Western Hemisphere* was published in 1937.

Ray and I were surprised to learn that William Berryman Scott traveled with a paleontologically inclined artist named R. Bruce Horsfall. Now largely forgotten, Horsfall's images of the extinct mammals of North America are some of the finest ever drawn. The scientist-artist duo is a trope that Ray and I originally thought was novel, but the more we read, the more we realized that we were just reliving an old tradition of natural history.

Ray was nuts about uintatheres. He naturally gravitates to the unusual or spectacular, and the knobby-headed uintathere had long been a favorite. Uintatheres first originated in Asia and migrated to North America via the Bering Land Bridge. They flourished here through the middle years of the Eocene before becoming extinct. Uintatheres were big beasts, the size of rhinos but with incredibly wide pelvic bones; in other words, they were shaped like pears. These were the easy fossils to find because their big, weathered skeletons could be seen for miles. But the Bridger Basin is also rich in smaller mammals. And whereas one could easily find a uintathere from horseback, it took careful crawling to find the other critters.

Leidy found and described the skeleton of a monkey like primate that he called *Notharctus*. Fossil turtles and crocodiles were common. And the most abundant mammal was the *Hyopsodus*, or tube sheep, which are so abundant in the Bighorn Basin. Like the Bighorn Basin, the Bridger Basin has remained a mecca for fossil mammalogists for the last century.

In the early 1990s, Richard Stucky of the Denver Museum began working in the Bridger Basin. By then, most of the easy-to-find fossils had been collected, so he concentrated on the smaller stuff. I joined the crew for two different seasons and was amazed by the number of crushed fossil turtles that littered the landscape. In one place we found a layer of crushed turtles nearly a meter thick. This interbedded mud and turtle lasagna really made me think about why turtles make such extraordinarily common fossils.

It's likely that when you're searching for fossil bone in the American West, the first fossil you find will be part of a poor deceased turtle. The more I thought about it, the more I realized that turtles are the perfect animal for fossilization. First, they live in ponds, lakes, and streams: places that are always accumulating mud and sand. Second, the turtle's shell is like a heavy box that sinks to the bottom of the pond when the animal dies. It dawned on me that turtles wear their own coffins and live in their own graveyard. Looking at these piles of crushed turtles,

A stack of Eocene turtles.

I couldn't help but think of Dr. Seuss and Yertle the Turtle, the king of the turtles who stacked all of his subjects in a huge pile so he could climb to the top and see all that he owned. And here we were, 50 million years later, looking at the fossilized remains of Yertle's folly.

Ray and I continued east on Interstate 80 across the north side of Bridger Basin and stopped at the original Little America truck stop for soft ice cream and to call my mom. A few miles later, we rolled down the hill into Green River and saw the majestic cliffs of the Green River Formation. These same cliffs had inspired 34-year-old Thomas Moran when he stepped off the train from Cheyenne in 1871. He fell in love with this view and sketched and painted it many times. His classic image, *The Cliffs of Green River,* was painted in 1874, and he was still painting versions 44 years later. Also here was the spot where on May 24, 1869, John Wesley Powell and his brave crew launched wooden boats that would eventually carry them through the unknown Grand Canyon.

We had an appointment to meet Dave Love's son Charlie in Rock Springs. Any son of Love's is bound to be interesting, and Charlie delivered in spades. Part geologist, part archaeologist, Charlie has made a career at Rock Springs Community College, where, oddly, he researched the giant stone statues of Easter Island in the South Pacific. He was trying to solve the age-old problem of how the Easter Islanders moved their monuments. His approach was pure small-town common sense with a dose of big

think. He began by building full-scale concrete models of the statues. Then, with rolling logs,, ropes, and Rock Springs students, Charlie showed how easy it was to move the massive stones. He had parked one of the giant concrete heads out behind the college, and I parked Big Blue next to it. Ray was impressed by the absurdity of it all.

Charlie knows the power of fossils, and he's turned an otherwise obscure community college into a flashy destination museum by working with a local welder to mount stylish and acrobatic casts of dinosaurs around the campus. The student lounge features a dynamically posed *Tyrannosaurus rex,* and the hallways are lined with a giant Green River turtle, a huge Cretaceous Kansas fish, a graceful Colorado plesiosaur, and a variety of dinosaurs. The lawns around the school support huge boulders from the Wind River Mountains, Charlie's attempt to bring the older rocks from the mountains' flanks to Rock Springs, where lazy geology students bump into them on their way to class.

Rock Springs is situated at the western edge of a geological structure called the Rock Springs Uplift. Like a smaller version of South Dakota's Black Hills, the uplift is an oval structure with the oldest rock layers in the center. Unlike the Black Hills, which expose Precambrian

basement rock at their core, the center of the Rock Springs Uplift exposes Late Cretaceous marine shale. The flanks of the uplift are formed by interbedded marine shale and coastal sandstone deposits, and the town of Rock Springs sits on the sandstone and coal of the Rock Springs Formation, an 80-million-year-old complex of beaches and swamps. As a result, Rock Springs has a long history as a coal miners' town; an old but surprisingly stylish neon sign down by the tracks still advertises the virtues of Rock Springs coal. The old downtown is a trip back to a time when Rock Springs was a coal-mining town on the Union Pacific line. Now the trains roll through every hour or so, but they no longer stop, and the classic old downtown buildings are occupied by tanning salons and junk shops.

A few blocks away are the archaeological ruins of a rock shop from my childhood. Den's Petrified Critters is a building that looks like it was made from log mill scrap. Ray and I peered through the broken glass and saw the remains of a once-thriving business of fossil fish. Doors hung lazily open on a couple of sheds in the backyard, and shards of fossils mixed with weeds along the battered wooden sidewalk. I remembered Den as a crusty fossil dealer known for using paint to enhance the charms of the fossil fish he sold. From all accounts, it looked like Den had left town. A few blocks away, at the slightly less dilapidated and still inhabited Tynsky Rock Shop, we learned that Den had headed to Arizona one autumn and never returned.

Tynsky is a big name in the fossil fish fields of southwestern Wyoming, but Tynsky namesakes were absent from this operation. The shop had traded owners a few times, and none of them had bothered to change the sign. The yard around the shop was piled deeply with chunks of petrified wood and *Turritella* agate. The owners, with an eye on the bottom line, had also branched out into aromatherapy, giving me yet another reason to mourn the decline and fall of the great American rock shop.

There was a time when almost every small town in the American West had a rock shop or two, evidence of the dual opportunities of the prospector and the treasure hunter. All you had to do was load up the camper, go out into the desert, and collect rocks. Somebody would eventually stop by to buy them. As a kid, I used to pore over the pages of *Lapidary Journal* and *Rock and Gem*. These magazines always included bad maps to remote spots where you could find your own rocks and minerals. In every city in the West, there was a network of rock and mineral clubs that would meet monthly to compare their finds and handiwork.

Many of these groups still exist, and they host annual gem and mineral shows. The largest by far is the February monster of a show in Tucson. All across the Rocky Mountain West, when a rock enthusiast says he's "going to Tucson," you know exactly what he means. It seems that everyone in the world with a cool rock or two loads up a van and heads to southern Arizona in February. Including everything from sprawling lobbies full of amethyst geodes to a couple of guys sitting quietly in a hotel room with a fossilized elephant skeleton, the Tucson show is one overwhelmingly weird nosedive into the world of the rock, gem, and fossil obsessed. Ray and I had "done Tucson" a couple of times, so we shared memories of the big event. In a lot of ways it's like the old mountain man rendezvous. Crusty old diggers come pouring out of the hills to sell the goods they've accumulated over a hard year and whoop it up with the similarly inclined. I suppose the addition of crystal healers, high-end jewelers, museum curators, schoolkids, and wealthy collectors makes the Tucson show an imperfect analogy. We decided that once again, we'd meet in Tucson in the upcoming February.

11
THE RAGING UINTATHERES OF VERNAL

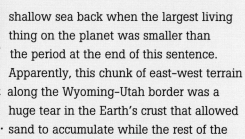

There aren't many paved roads south of Rock Springs. This region contains the Sand Wash and Washakie basins, areas revered by paleontologists for their fossil mammals. In the middle of this 5,000-square-mile forgotten quarter is a haunting maze of fossil-filled badlands known as Adobe Town. About 60 miles south of Rock Springs, as the road started to head downhill, we rounded a corner and saw a deep basin of brick-red badlands.

We had entered Utah. Almost immediately, the geology began to play ball. As we approached the Uinta Mountains, the layered strata older than the Wasatch Formation had been tilted up so that it dipped north and exposed as a series of east-west ridges, each older than the previous one as we headed south. By the time we reached Dutch John, we'd driven all the way from the Eocene back into the Precambrian. We paused at the dam to glance at Flaming Gorge Reservoir, which was buzzing with ski boats, Jet Skis, and small cruisers. We spotted a geologic road sign that claimed to explain the local geology, and eagerly pulled over to inspect it. The sign was pretty, but the geology was so miserably explained it virtually guaranteed that curious travelers would walk away confused and uninterested. We harrumphed loudly and jumped back into Big Blue for the steep climb up the Uinta Mountains.

The Uintas are the odd mountains out when it comes to the Rockies. To start with, they're oriented east-west. Then, unlike all the rest of the Rockies, which are cored by Precambrian crystalline rocks, the Uintas are cored by Precambrian sedimentary rocks. The Precambrian Uinta Mountain Group is a nearly three-mile-thick pile of layered fossil-free sandstone that formed at the bottom of a

shallow sea back when the largest living thing on the planet was smaller than the period at the end of this sentence. Apparently, this chunk of east-west terrain along the Wyoming-Utah border was a huge tear in the Earth's crust that allowed sand to accumulate while the rest of the region was suffering the ravages of E-World.

Downstream from the dam, the Green River winds lazily through Brown's Park, a rolling, drab landscape floored with Miocene sediments. Then the river does that amazing thing the Rocky Mountain rivers do: it flows directly into a mountain range, once again demonstrating that the river is older than the mountains. At the Gates of Ladore, the Green flows into a 2,000-foot-deep canyon whose walls are composed of layer after layer after layer of 1.6-billion-year-old Uinta Mountain Group sandstone. If you launch a raft at Ladore and float the Green, you will literally float through a mountain range. After crossing the center point of the range, the bedrock begins to dip to the south, so the river climbs up through the layered rocks while still flowing downhill.

After a long grind up through a thick pine and cedar forest, we crested onto a high plateau covered with patches of lodgepole pines and aspen. Once the road started downhill, we began to notice highway signs that pointed out the geological formations. The first few signs were odd, because they'd been placed in the woods, where it was impossible to see any rocks. We had pity for the hapless geological tourists who tried to make heads or tails of these wretched signs.

Then, as the road steepened, the rocks came into view, and we rolled down the south face of the range and climbed into younger and younger rock layers, because the tilt of the layers was steeper than the grade of the road. The main descent occurred over half a dozen switchbacks that snaked along the margin of an active phosphate mine. We realized as we drove that this was the Permian Park City Formation, a lateral equivalent to Ray's beloved Phosphoria Formation. This meant that the miners, unbeknownst to themselves, might be having casual encounters with whorl-toothed, or rather buzz-toothed, sharks. To my knowledge, no one has ever found one of these sharks in this mine, but they must be there, just like they're in the mines near Afton, Wyoming.

At the bottom of the hill, just past the mine entrance, the highway flattened out and rock layer after layer came splendidly into view. As I drove past the formations, I chanted the strange and evocative rock-reciting poetry of a driving geologist to my bemused artist friend. Ray cranked James Brown on the CD player, and I belted out the names of the formations as we passed them: "Dinwoody, Moenkopi, Shinarump, Chinle, Nugget, Carmel, Entrada, Curtis, Morrison, Cedar Mountain, Dakota, Mowry, Frontier, Mancos." In the 20 miles between the mine and the edge of the town of Vernal, we drove through 13 different Mesozoic formations, from the brick-red 225-million-year-old Dinwoody to the 80-million-year-old monotonous dark gray Mancos Shale. We stopped several times to look at the layers, finding bits of fossil fish in the black flaky shale of a fresh Mancos road cut and feeling like the funk from the CD had called the fossils from the hill.

We gave up when the sun went down. Hungry and happy, we rolled into the dinosaur-obsessed town of Vernal and got a room at the Best Western Dinosaur Inn, where we found small plastic dinosaurs next to the complimentary toiletries. After a few poolside gin and tonics and a bona fide ranch-sized meal at the 7-11 Ranch Café, we made our way back to the hotel and slumped into the hot tub.

Apparently, a bunch of local high-school kids knew the code to the hot tub room, and we were soon joined by a trio of teenage Vernalites. After our attempt to change the school mascot in Afton, Wyoming, to a whorl-toothed shark, we thought that we'd have a go at changing the Uintah High School mascot to the Raging Uintatheres. It's tough to convince someone to give up their team mascot, but we figured that the present name, the Uintah Utes, wasn't politically correct anyway, so Ray broached the idea with the kids. They looked at us like we were completely out of our gourds and moved over to the other side of the pool in embarrassed silence.

Vernal is a remote and unremarkable town made exquisite by its geologic setting. Located just south of the Uinta Mountains on the northern margin of the Uinta Basin, it lies at the center of a paleontological paradise. This treasure was initially discovered by O. C. Marsh, who focused on the flat-lying Eocene rocks in the middle of the basin and was handsomely rewarded with a hoard of Eocene mammals and a jackpot of uintatheres. But it was Andrew Carnegie's man Earl Douglass who really made the place famous. In 1909, he found a string of giant vertebrae on a tilted ridge of Morrison Formation sandstone high above the north bank of the Green River where it spews out of the Uintas at Split Mountain. Douglass excavated his way into the tough sandstone, and soon the ridge was a veritable dinosaur mine, producing splendid skeletons of *Apatosaurus, Camarasaurus,*

VERNAL HIGH SCHOOL
RAGING UINTATHERES

Ernest Untermann (left) and Ray Troll (right) with their prehistoric muses.

Diplodocus, *Stegosaurus*, and *Camptosaurus*. His efforts stocked the Carnegie Museum with dinosaurs, and there were enough left over for the Smithsonian and the University of Utah. After all the smoke had cleared, nearly 40 skeletons had been chipped and blasted out of the ridge, and parts of dozens remained embedded in the mountain wall.

In 1915, the place was named Dinosaur National Monument. Douglass built a cabin on the banks of the Green River and spent the rest of his life collecting Jurassic dinosaurs and Eocene mammals. In 1958, a building was constructed over the remnants of the Douglass Quarry. Regardless of the huge number of skeletons that have been removed, this remains the best place in the world to see dinosaur bones in place, a virtual gateway to the Jurassic. In a response to the dinomania that swept the region, the state of Utah decided to open a small museum of its own in Vernal. The Utah Field House of Natural History State Park opened its doors in 1945. At the time of our visit, the state of Utah was planning a major renovation to the old building. I'd been retained by a Seattle design firm to act as the paleontologist for the renovation, so I'd recently made several trips to Vernal and was familiar with the collections and the museum. The old building was still intact and open for business, and I was anxious to show Ray the highlights of this classic old joint.

We woke to a nearly perfect day that happened to be the Fourth of July and ate a substantial breakfast. We finished just in time for the parade down Main Street.

Vernal is an all-American town, but it's also a dinosaur town, so the parade combined the usual rodeo queens and veterans in convertibles with a whole host of giant papier-mâché dinosaurs. We thought they were missing a beat by not exploiting the uintathere theme.

The sun was starting to broil and I was raring to show Ray the treasures of the Field House, so we walked half a block from the Dinosaur Inn to the squat, square brick building. The entry foyer was nondescript, but it contained five notable objects: two petrified trees, two paintings, and a stone transom inscribed with the words "A knowledge of the past makes understandable the present and serves as a guide to future." It was my kind of place.

The paintings were fantastic. One showed a Triassic phytosaur lounging in a compelling and realistic yet strangely Dr. Suessian forest. In my mind, most prehistoric landscape paintings are ruined by the irresistible pull of the present. Who can blame the artist who paints familiar landscapes populated by extinct animals? Artists, after all, are animals of the modern landscape themselves, and there are very few good references to work from when reconstituting extinct ecosystems. Nonetheless, I almost always feel that these paintings look too modern, too "today." But not these strange pastel-colored canvases hanging freely at the Field House entry. We had entered the world of Untermann.

Ernest Untermann was a friend of Jack London and translated London's work into German. He was an ardent Communist who also translated Marx into English. Born

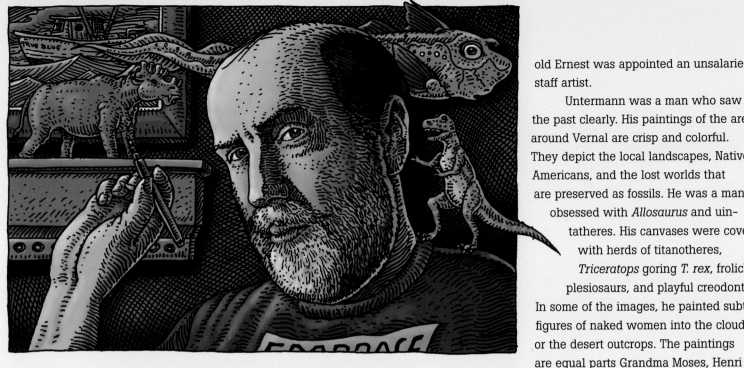

old Ernest was appointed an unsalaried staff artist.

Untermann was a man who saw the past clearly. His paintings of the area around Vernal are crisp and colorful. They depict the local landscapes, Native Americans, and the lost worlds that are preserved as fossils. He was a man obsessed with *Allosaurus* and uintatheres. His canvases were covered with herds of titanotheres, *Triceratops* goring *T. rex*, frolicking plesiosaurs, and playful creodonts. In some of the images, he painted subtle figures of naked women into the clouds or the desert outcrops. The paintings are equal parts Grandma Moses, Henri Rousseau, and ribald paleontologist. In the nearly 20 years between his move to Vernal and the day he died at the age of 92 in 1956, Untermann painted hundreds of canvases. These paintings had spread around town, showing up in restaurants and private collections, some changing hands at yard sales, but most of the vast body of work staying where he worked, stashed in the fossil museum.

While Ernest painted, Getty and Billie built the museum into a local treasure. They fanned the local dinosaur frenzy and oversaw the construction of a dinosaur garden populated by huge fiberglass dinosaurs of dubious proportions. They even acquired a cast of *Diplodocus carnegii*, the animal from Sheep Creek, Wyoming, that Carnegie had sent around the world. For years, the big concrete *Diplodocus* graced the front yard of the Field House, and Vernal had the opportunity to say, "London, Paris, Rome, Buenos Aires, Vernal." Of course, it didn't think to say that, but it could have.

While dozens of the paintings were hanging on the walls of the museum, there were scores more gathering dust in the museum's antiquated attic, an odd space that was no more than five feet tall. The museum's curator, Sue Ann Bilbey, led us to them. We made our way up

in Germany in 1864, Untermann went to sea on American sailing ships at the age of 17 and had the misfortune but romantic good luck to be twice shipwrecked in the South Seas. After the second wreck, he drifted alone in an open boat for 21 days before washing ashore on an island. He lived there with the locals for more than a year before being rescued. By the time he was 29, Untermann had come ashore, studied geology and paleontology, and was working in the Rockies, briefly visiting Vernal in 1919. This was followed by stints in the 1920s and 1930s collecting zoological specimens for scientific supply houses in Brazil and East Africa. His interests in paleontology and animals led him to pursue art. He studied at the Chicago Art Institute and later under a painter named William Heine, who was famous for painting grand battle panoramas. Tireless and versatile, Untermann embroiled himself in politics on one hand and animals and art on the other, ending up as the director of the Milwaukee Zoo. In 1940, after a controversial tenure no doubt soured by his politics, he retired from his position and, in his 75th year, moved to Vernal to live near his son George, also known as Getty, who was a ranger at Dinosaur National Monument. In 1945, when the Field House opened its doors, Getty was appointed its first director, Getty's wife, Billie, was named staff scientist, and

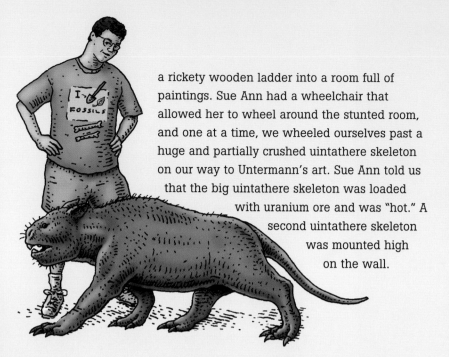

a rickety wooden ladder into a room full of paintings. Sue Ann had a wheelchair that allowed her to wheel around the stunted room, and one at a time, we wheeled ourselves past a huge and partially crushed uintathere skeleton on our way to Untermann's art. Sue Ann told us that the big uintathere skeleton was loaded with uranium ore and was "hot." A second uintathere skeleton was mounted high on the wall.

Kirk and a taeniodont, the bizarre Eocene mammal that seems to be all about teeth.

Apparently this one, a papier-mâché model, had been purchased from Ward's Scientific Company sometime in the distant past. The paper was shredded currency from the mint. Then came the paintings. Ray was in heaven, trying to channel Untermann through his images of Eocene mammals. We found an amazing self-portrait of the old commie with a tiny uintathere posed regally above one shoulder and a snarling Allosaurus gnashing its teeth over the other. "Oh my God, it's like a Van Gogh with two dueling muses," Ray murmured. He had clearly met his own muse, and we spent the afternoon pulling one masterpiece after another from the rickety rope contraption that held the frames. Funky old paintings, radioactive uintatheres, and uintatheres made of cash: this place was too good to be true.

We returned to the main part of the Field House to admire the fossils on display. In one case there was the skull and jaws of an animal that would come to captivate us both. It was a stylinodontine taeniodont, a beast whose lineage appeared around 65 million years ago and lasted until 42 million years ago when it checked out for good. Taeniodonts are leopard-shaped mammals with massive claws on front and back feet. Their jaws are crammed full of pluglike teeth. The result is a buck-toothed wombat with a catlike tail. Ray goes for the creatures that no one

has ever reconstructed, or at least reconstructed well, and the taeniodonts would become an obsession.

Later, in the cool of the evening, we wandered down Main Street to the middle of town. All around us the locals were lining up lawn chairs in vacant lots, anticipating the fireworks. There was a half-block-long white building on the south side of the street that I wanted to check out. I'd been there once before back in the early '90s when the building had housed a truly amazing fossil shop. I bought a beautiful fossil crab from the Cretaceous of Tennessee and a stunning finned ammonite from the chalk beds near Fort Worth; both prizes of the *Prehistoric Journey* exhibit at the Denver Museum of Nature & Science. The proprietor was a tough-talking woman named Lace Honert. Based on the fantastic fossils I saw in her shop that day, it was clear that Lace was a bona fide digger. I was curious if she was still around.

The storefront windows were covered with white paper, and the shop was clearly no longer in business, yet the wonderful sign on the side of the building still displayed the name of the store, "Remains To Be Seen," high above the street. I knocked on the door. Nothing. I pounded. This time Lace materialized and invited us in. She had friends over, and they were sitting on the big, flat tar roof drinking beers and waiting for the fireworks to start. We climbed over her toilet and wriggled out the bathroom window to join them. Ray and I told Lace that we were driving around the West looking for the best fossil stories. She lifted an eyebrow and squinted at us. Sizing us up, she handed out beers and said, "Well, I've got a few stories for you."

In the off-season (the non-digging season, that is), Lace lives in Texas. She's married to a guy named Jim who was once Pete Larson's partner. Lace told us

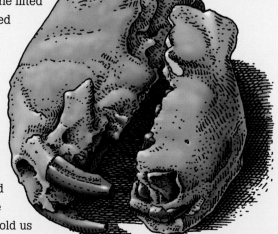

STYLINODON SKULL

that she decided to marry Jim because of his "localities." Ray said, "Localities?" "Localities," Lace replied, clearly annoyed that Ray didn't understand she had chosen her mate because of his knowledge of where to find fossils.

Somehow Lace had learned about Untermann and had driven out to Vernal with the hope of acquiring one of his paintings. Instead, she ended up buying the building on Main Street and a piece of fossil-rich land near the town of Dinosaur, Colorado, where she runs her very own dinosaur quarry. She often works alone, wrestling giant

dinosaur leg bones out of the ground when the need arises. She remains the only woman I know who owns her own backhoe. Tellingly, her female fashion sense is intact but a bit warped. The Imelda Marcos of the fossil-hunting set, Lace owns more than 30 pairs of field boots.

After a few more beers, Lace could see that what Ray and I were really after was an adventure of some sort. She started talking about a truly fantastic fossil site that she had stumbled across while hiking near Dinosaur National Monument. As the fireworks started up, she promised to take us there the next day.

The next morning dawned a sizzler. Lace pulled up in a very large truck, and we rolled down the road from Vernal to Jensen, a long commercial stretch festooned with vintage commercial dinosaur sculptures. After entering the monument, we crossed the Green River and drove until the road changed to red dirt. Lace parked at the base of a large slope, and we climbed a couple of hundred feet up to the base of a huge sandstone cliff. Because I'm a big guy and I like climbing hills, I moved faster than Lace and Ray and was standing at the base of the cliff when they huffed into view. Just then, I realized that the rock face was covered with subtle petroglyphs.

Lace pointed to my right where a huge chunk of the cliff had broken away, creating a giant fissure. The opening was less than a yard wide, but the crack was easily a hundred feet high. I stepped into the opening and stopped, realizing that spooky life-sized guardian figures were engraved on both facing walls.

I know that a lot of people get the willies around big petroglyphs. Ray is one of them. He could feel the power of the art, and he started to freak a little. Lace smiled and said, "See, cool place," then sauntered up the crack. Ray noticed that there were giant spirals etched on either side of the fissure. "The Indians must've dug whorl-toothed

sharks too," he said as he stepped into the big hallway of rock.

The floor of the crack sloped steadily up, and within about 100 yards we were at the top of the cliff standing at the edge of a stunted juniper forest. Lace took off and we followed her for at least a mile, dodging trees and sandstone slabs until we came to a small canyon. We lowered ourselves down the rock walls and followed them up valley for a couple of hundred yards to a 10-foot overhang. Lace hung back and said, "Check it out."

I crawled under the overhang and looked up. There, above my head, projecting down from the flat sandstone surface, were the polygonal projections of fossil mud cracks. Scattered across the surface were a medley of small tracks, some three-toed, others four. The mud between the tracks was a lovely bright green, and the rock itself was red. I looked down and saw the sandstone slab that had fallen to create the overhang. Its surface was the mirror image of the surface from which it had fallen. The rock was just so beautiful, the colors, the polygons, the tracks. Lace really had known what we were after.

The three of us sat there, imagining the little Jurassic dinosaurs skittering across a drying lake bed, a moment from 180 million years ago frozen in time. I tried to picture the rest of their world. The cliff was part of the

Fossil leaves are like potato chips: you can't have just one. These *Macginitiea* leaves were recovered from a 47-million-year-old lake bed.

Nugget Sandstone, a Jurassic formation known because it's largely composed of lithified sand dunes. It's thought to represent a Sahara-like sand sea that covered much of the Four Corners region around 180 million years ago. Few complete skeletons are known from the Nugget, and fossil plants are even more rare, although the occasional fossil tree trunk does show up. These tracks, perfect as they were, represented an incomplete message from a lost time. Truly impressed, we climbed back out of the canyon and headed back into the 21st century.

Although we'd all been there before, we decided we shouldn't pass up the opportunity to see the giant quarry wall at Dinosaur National Monument. The fact is, Earl Douglass really did discover one of the greatest fossil sites in the entire world. We drove up the winding road from the valley of the Green River, back through the tilted layers of time, and joined the throngs of tourists already streaming into the large building that enclosed a dinosaur-studded rock wall. For years, park service employees wearing campy white jumpsuits and hard hats crawled around on the bone-infested surface and chipped away at the hard rock, exposing more bones and posing for funky postcard photographs. The result is a Jurassic sandbar frozen in time and laid out for our viewing pleasure. Parts of dozens of incomplete dinosaurs jumble the surface: a leg here, a section of tail there, and near the top of the wall a length

of neck terminated by a nearly perfect *Camarasaurus* skull. The Morrison Formation is known for its diversity of long-necked sauropods, and it's crazy to see how these 50- to 80-foot-long beasts fell apart. They fossilized in segments of tail vertebra, neck bones, and logjams of legs.

Despite the richness of fossils on the wall, the setting that caused the fossils to be there is not immediately clear. The signage suggests that it was the lithified and tilted sandbar of a giant river. Not implausible, but why are there so many skeletons? It still baffles me. The discovery of cracks in the building's foundation in 2006 caused the National Park Service to close this magnificent attraction. Federal funds were quickly found, and a new quarry building opened in 2011.

After the visit, we dropped Lace in Vernal and headed due south to the nearly abandoned town of Bonanza. Along the way, the road from Vernal cuts across the middle of the Uinta Basin and rolls through the drab gray badlands of the Uinta Formation. Both the Uinta and the overlying red beds of the Duchesne Formation are known for their larder of Eocene mammals. As we continued south, the rolling badlands gave way to sagebrush flats with occasional sandstone hills. We crested a final hill and rolled into Bonanza.

Bonanza is more of a mine than a town. The mineral of choice is a bizarre hydrocarbon known as gilsonite, used for paints and plastics. The strange thing about this shiny black asphalt-like substance is that it occurs in long vertical veins that track across the landscape in a northwest-southeast orientation. The Bonanza seam was a whopper, nearly 20 feet thick, and it stretched for miles. During the heyday of gilsonite mining, workers dug straight down into the seams, building a wooden lattice to keep the resulting trench open. Eventually, some of these narrow trenches ran for miles. The roads around Bonanza cross these immense slits, and for the casual weekend driver, the scars beg nothing but questions.

South of Bonanza, we plunged down a valley into the canyon of the White River, then up a long sandstone ridge and back down into the valley of Evacuation Creek. Once we dropped to creek level, we descended from the Uinta Formation to the top of the Green River Formation and back into the remains of the old Eocene lakes.

Back in 1991, a Vernalite named Bruce Handley invited me out to dig fossil leaves near Evacuation Creek. He'd perfected the art of popping up big slabs of ancient lake bed and revealing fantastic leaf fossils. This is no easy trick. The hard shale of the Parachute Creek Member of the Green River Formation is finicky stuff. It splits readily into thin sheets after it has weathered for a while, but the freshly exposed rock is hard as Hades and breaks into rounded, conchoidal chips rather than perfectly flat surfaces. Bruce figured out how to judge the slope of the hill just right so that he could pop up plywood-sized sheets that had weathered to just the right degree of split. The fossils that he found were exquisite leaves, flowers, and

Hell Hole Canyon near Bonanza is a superb exposure of the Green River Formation.

India Wood's *Allosaurus* skeleton.

insects, preserved as a sweet carbon black on a creamy matrix. Using my adage that "you can't find a big leaf on a small rock," Bruce mastered the art of pulling table-sized slabs from the earth and finding whole fossil branches.

During the planning for the new Vernal Field House Natural History Museum, I got the idea that it would be cool to make an entire museum wall out of slabs of this stone. I wanted to scout an appropriate site to quarry the fossils. Bonanza has been discovered by fossil hounds, and the sage-covered slopes are pocked by quarries made by their efforts. Ray and I walked around one of my old quarries and split a few slabs. Fossils were pretty easy to find, and it's not too hard to imagine that this was once the bottom of a Lake Erie–sized lake. Later that year, I returned with a team of 30 volunteers to quarry enough fossil slabs to fill out a 300 square feet of wall with rectangular leaf-bearing slabs from the fossiliferous lake bed, which now hang on the museum wall in Vernal. This project turned out to be more than I bargained for since we had to quarry the fossils and saw them into rectangles in the field. My volunteer corps started to flag, and someone suggested that I reach out to the local sheriff for help. I did and he loaned me a couple of prisoners from

the Vernal jail. These guys showed up wearing the black-and-white striped shirts and pants that you used to see in old television shows about prison. The sheriff called them zebras, and those zebras helped us get the job done.

The Utah Field House of Natural History State Park Museum (yes, that is its name) opened in a brand-new building in Vernal in 2004. When viewed from above, the building is in the shape of a nautilus shell. I worked with designer Allison Craig Sundine and Karen Krieger, the project manager from Utah State Parks, to design the exhibits inside the new museum. We were sitting in a restaurant drinking coffee when I sketched a spiral on a napkin and suggested that it would be cool to shape the building like an ammonite. Mission accomplished!

The museum contains an Eocene diorama that shows what it would have looked like in the Uinta basin 48 million years ago when basin was a great lake. We had contracted artist Gary Staab to build life-size models of several Eocene animals. He outdid himself and created two uintatheres, a taeniodont, a bushbaby-like primate, a protoreodon, a turtle, a varanid lizard, a boa constrictor, and a dragonfly. When it came time to load in the models, he discovered that the pear-shaped uintatheres had backsides that were too wide to fit though the door. Fortunately, it was a near miss, and he was able to file off some excess hind quarters and the two giant beasts now grace a beautiful diorama. The next room over has the uintathere skeleton made of cash and the giant wall of fossil leaves. A few years later, Vernal added a collection facility that makes it easy to access the museum's amazing collection. I visited the museum in 2022 and was delighted to see dozens of Untermann paintings hanging salon style on the walls of the giant collection warehouse.

On the day Ray and I were there, it was still plenty hot and I wasn't up for a major quarry effort. Instead, we drove up a two-track to the top of a nearby ridge and peered off the other side. One of the great things about the American West is how you can find your own vistas and think, because of the remoteness and the emptiness, that you're the first person to ever see them. The view into Hell Hole Canyon, which lay before us, is like this: smooth and nondescript on the west side, but on the east one of the great natural outcrops in the world. Nearly a thousand feet deep, the canyon walls are composed entirely of thinly bedded Green River Shale.

If there's one place on Earth where the metaphor "pages of time" is a reality, this is it. When ancient Lake Uinta stood at this spot, mud settling to the lake bottom accumulated steadily enough to help overburden the Earth's crust. The result was that the bottom of the lake sank even as it was accumulating layers of mud. This went on for millennia, and the result was a 1,000-foot pile of mud at the bottom of a 100-foot-deep lake. Hell's Hole slices through this huge fossil phone book. It's a hard spot to find but well worth the effort.

After winding our way back to Bonanza, we headed out of the Uinta Basin toward Rangely, Colorado, passing through the odd little town of Dinosaur. This is a place that has tried to make dinosaurs pay, but there just isn't enough real traffic through this dusty corner of Colorado for anyone to break even, much less make a buck. I must admit that I'm touched by the genuine Dinosaur Cemetery, which is full of deceased human residents, but Brontosaurus Street doesn't quite do it for me. The city park does have a splendid pair of incredibly ugly dinosaur sculptures: an overhorned *Triceratops* and an extra-bony bone-headed *Pachycephalosaurus.* Ray thinks highly of such roadside works of art and posed me for a snapshot by these battered monuments. "It's pretty cool that some old guy out in the middle of nowhere would be so moved by dreams of prehistory that he was driven to resurrect these beasts. Think of all the bags of cement," Ray mused.

East of Dinosaur, Highway 40 is a long empty road that eventually leads to Craig and Steamboat Springs, but the first 60 miles contain absolutely nothing save a few grassed-over sand dunes. One of Colorado's more colorful dinosaur discoveries happened in this barren corner of the state. In 1979, a single mother dropped her only daughter off at a friend's ranch for a summer. The girl, 13-year-old India Wood, was an independent and brainy youngster who liked to ride horses and had a lot of time on her hands. At some point during that summer,

India found some black dinosaur bones weathering out of an exposure of the Morrison Formation and started to dig into the hill. With the help of the ranchers, who knew a fossil when they saw one, and a few library books, India educated herself in the ways of paleontology and set about excavating a skeleton that she correctly identified as an *Allosaurus.* Over the course of the next three years, India pulled more than half of the skeleton out of its muddy matrix and carefully hauled the bones back to her home in Colorado Springs. By the time India was 16, her mom was beginning to worry about the volume of rock in her bedroom and suggested that she find a new home for her dinosaur. That's when India called Don Lindsey at the Denver Museum and said, "I've got an *Allosaurus* in my bedroom, do you want to see it?" Much to Lindsey's credit, he headed down to Colorado Springs with the museum's videographer to check out India's claim. The video footage, still in the museum's archive, shows a young girl's room full of plastic horses and a pink bedspread. Then young India starts to pull box after box of beautifully collected dinosaur bones from beneath her bed.

The museum hired her on the spot, and for the next few summers, India guided a museum crew back to the site to collect the rest of the skeleton. At the end of the day, Lindsey wasn't a very good mentor, telling India that there was no future for her in paleontology. She took him at his word and went off to get an MBA. Thirteen years later, the *Allosaurus* was pulled from a dusty storeroom and installed next to the Colorado state fossil *Stegosaurus* as the centerpiece of Denver's *Prehistoric Journey* exhibit.

Acting on some old information, I tracked India down in Cambridge, Massachusetts, where she was living with her husband and two kids, and flew her to Denver for the opening of the exhibit. India and I became friends, and eventually, she and her family moved back to Boulder. A few years ago, I finally had the chance to visit the site of her childhood discovery with her and stand in the hole where she came of age digging her own *Allosaurus*.

On that memorable visit, I drove out from Denver in my Saab and followed some tortured directions that took me down several long dirt roads to a ranch house. The

place was a museum in its own right, as the ranchers had collected all manner of hides, heads, fossils, rocks, artifacts, and bones, filling their living room. India cooked me a fantastic steak, then we drove out along a long valley and camped near a juniper forest. The next morning, we woke up and hiked over to the spot where India found her animal. More bones were exposed nearby, and India showed me where she'd found a *Camara-saurus* vertebra the day before. Nearby, a truly bizarre excavation was underway. A group of Texas creationists were busy excavating a *Stegosaurus*. It wasn't clear whether they knew they were proving themselves wrong or if they thought they were testing a theory about the packing of large animals onto Noah's ark. I could tell that she was a bit unnerved to be back at the site of her childhood discovery. My presence there seemed like an unmerited intimacy, and the creationist dig seemed sacrilegious in an anti-in-tellectual way. I quietly cursed Lindsey for dissuading this talented woman from entering the field of paleontology.

A year later, Ray and I passed the turnoff on the dusty ranch road to India's *Allosaurus*, and I told him about the creationist dig. He thought for a moment and said, "What blows my mind is that evolution is the biggest puzzle mankind has ever solved, and it took generations to do it. Everything, and I mean everything, falls into place when you perceive the world through evolutionary eyes. How can these guys not see the logic of what they're denying?" I agreed.

It was nearing dusk when we hit Rangely for a family-style Italian meal of spaghetti and meatballs at Mangilino's Diner. It was one of those nights when we hadn't really decided where to sleep, and it was still light after our big meal, so we headed down Highway 139 with the thought that we might camp on Douglas Pass. The road south from Rangely runs straight south through increasingly high cliffs of Mesaverde Sandstone. Several times in the waning evening light, I spotted what looked to be dinosaur footprints in sandstone outcrops along the side of the road.

We inspected a few by the headlights, but never really convinced ourselves that we'd found a good one.

Douglas Pass is at an elevation of over 8,000 feet. A spruce and aspen forest closed around us as we gained elevation and lost the last of the daylight. At the top of the pass, we turned off the pavement to the east. I shifted Blue into low gear and slowly climbed a steep but oddly wide gravel road that gained several hundred feet before leveling off on a forested ridgeline. We drove down the ridge for about a mile, then I pulled Big Blue off the road into a little meadow. It was delightfully cool as we laid our sleeping bags out on the ground, and soon we were dreaming about finding the perfect Eocene leaf.

At dawn the next morning, we were awakened by the sounds of pots and pans banging nearby. Then, voices started discussing fossils. Inadvertently, we had camped within 40 yards of some other fossil diggers. I walked over to their car and recognized some of my own volunteers from the Denver Museum of Nature & Science. I'd completely forgotten that there was a big field trip to Douglas Pass over the July Fourth weekend. For years these guys had been asking me to join them, but other July Fourth events had always intervened. Now, with absolutely no intention what-soever, I had accidentally shown up for the dig. I cleared my throat and brazenly lied that I'd been planning to be here all along. A quick nod to Troll and a subtle blurry-eyed wink back assured me that he was in on the ruse. "Got any coffee?" were his first doleful words of the day.

Like the site near Evacuation Creek, Douglas Pass is an exposure of the Parachute Creek Member of the Green River Formation. This is one of the richest horizons of oil shale in world. Oil shale is a tight, hard rock that is black and smells of oil when it's fresh but is nearly white and splits into perfectly flat, thin sheets when weathered. There was a lot of interest in oil shale in the 1970s during the Middle East oil crisis, and a lot of money was spent gearing up to make the area around Rifle and Parachute, Colorado, into a major petroleum-producing region. Kuwait, Qatar, and Parachute—it had a nice ring to it. There certainly was enough of the stuff to get people whipped into a get-rich-quick frenzy. The oil shale formed as algae and other organics sank to the bottom of the Green River lakes way back in the Eocene. The lakes were huge, and in the Uinta Basin of Utah and the Piceance Basin of Colorado, the conditions were just right to bury a huge amount of shale.

The problem was that the shale didn't want to give up its oil. You could drill a well right into the middle of the oiliest, blackest, richest layer, lovingly called the Mahog-any Zone, and nothing would happen. No gushers, no Jed Clampett bubblin' crude, no Texas tea, nothin'. The rock was simply too tight; the oil was there, but they just couldn't get it out.

It was pretty frustrating to be sitting on the next Prudhoe Bay and have no way to make it pay. All manner of schemes were cooked up to make the shale yield its treasure. One thought was to mine the shale, crush it, and cook it so the oil would ooze out. Another was to drill a hole, then set off an explosion at the bottom of it, thereby fracturing the rock and releasing oil. In principle, these approaches worked, but they always cost lots more than the oil was worth. The culmination of effort and absurdity happened in 1969, when a consortium drilled a well in the Piceance Basin and lowered a thermonuclear device to a depth of 7,000 feet. They set the device off, Colorado's only nuclear explosion, and sat back and waited for the bubblin' crude. But it never came. The idea had been to create a huge underground cavity into which the oil would flow. The problem was that the bomb, known as the Rulison blast, fused the bedrock into a huge mass of glass. Probably a good thing, as it's not clear to me what you would do with radioactive petroleum. To this date, the state restricts drilling in this area to prevent inadvertent release of radioactivity.

Finally, in 1979, Shell Oil, the major player in the big boondoggle, pulled the plug and walked away from the oil shale boomtowns of Parachute and Rifle, leaving the economy to scramble for new foundation. Here and there you could still see small experimental operations where oil shale is baked to make road-grade asphalt, but, by and large, the exploitation of the oil shale will likely never happen.

In the meantime, paleonerds like Ray and me continued to flock to Douglas Pass. The most famous fossil site there is known as Radar Dome, because the FAA has one of its distinctive giant–golf ball microwave facilities located at the top of the large round hill. The hill is made of oil shale, and the Mahogany Zone is about 150 feet below the dome. The west side of the hill drops away precipitously. The rocks above the Mahogany Zone are full of gorgeous fossils, and the land is owned and managed by the BLM, so it's legal to collect reasonable amounts of plant and invertebrate fossils for noncommercial purposes.

It used to be that you could drive right to the dome and start digging, but then some clever person realized that having fossil-digging families milling around directly in front of the giant microwave transmitter wasn't the best thing for everybody's well-being, so they gated the road. The fossil diggers just moved their digs along the access road and around the back side of the hill.

At the Denver Museum of Nature & Science, we name our fossil sites as well as numbering them. Numbers are great for the computer database, but it's just easier for our brains to remember names. The museum name for the fossil quarry along the west side of this hill is "Da purdiest fossil site in the world," and the collection drawers at the museum bear this happy moniker. We named the site on a perfect summer day back in 1998 when we were popping up sled-sized sheets of half-inch-thick shale that were covered with extra-fine fossil leaves. It was a sunny day but not too hot, not too much overburden, no biting insects, no wind, a cliffside view that looked out west past the Utah border, a happy crew of good-lookin' young diggers—who could ask for more than that? We had a photographer from *U.S. News and World Report* with us that day. They published a photo and a small blurb about fossils, but I doubt that any of the millions of readers of that little piece had any idea of what a fine time they were missing.

The light rock and dark imprints in oil shale make for really pretty fossils, and the chance of finding something rare gives the site what we call "that Green River feeling." This feeling is the fairly legitimate hope that you might actually find something spectacular. It's like buying a lottery ticket, but with great odds. Yet unlike buying a lottery ticket, the process itself is lots of fun.

The bread-and-butter fossils of Douglas Pass are leaves and insects, common enough that people who've never dug before can expect to go home with boxes of them. The rare stuff comes in several forms. It can simply be a fossil that is extremely well preserved: a leaf that shows tiny veinlets in detail; an insect with a stinger still intact; or a fossil flower with perfectly preserved and obvious pistils, stamens, and anthers. Or it can be a whole fossil branch complete with attached leaves and sometimes even flowers. Every once in a great while, someone will find a small fish, but they tend to be guppy-like things. If you really want fossil fish, go back to the private quarries near Kemmerer, Wyoming.

The jackpot fossils at Douglas Pass are things such as whole moths and butterflies with the patterning of their wings preserved. Once, at the fossil show in Tucson, I was shown a hand-sized slab from Douglas Pass with a complete but flattened baby bird, its tiny pin feathers, bony feet, and pointy beak all clearly visible. But there's a rub. The BLM regulations say that it's legal to collect plant and invertebrate fossils for noncommercial uses. So you can't collect vertebrate fossils, and you can't sell or trade the plants and invertebrates you find. It's sort of like saying, "Play the lottery, but only keep the payouts that are worth less than a hundred bucks." The collector of the baby bird defied those rules, and another rare fossil was lost to science.

The group from Denver that Ray and I ran into were members of a group known as WIPS, short for the Western Interior Paleontological Society. WIPS is one of several fossil clubs around the country that gathers monthly to talk about the many joys of loving fossils. The trip leader was a determined man named Mike Graham who had made Douglas Pass the focus of his hobby life. Mike is a computer database whiz who has collected every scrap of science that has ever been written about the plants and insects of the Green River Formation. Mike and the BLM put their heads together and worked out a plan to allow

A classic Douglas Pass fossil frog-legged leaf beetle that is preserved so well you can see the original patterning of its wing covers. In 2021, this specimen was used to describe a new genus and species, *Pulchritudo attenboroughi*. The name means Attenborough's Beauty and it was named in honor of David Attenborough.

people to collect responsibly. At the end of each trip, Mike brings all of the unusual fossils to Denver, where we evaluate which pieces are so rare or scientifically significant that they belong in a museum. Those are transferred to the museum, and the others go back to the people who found them. In this way, the Denver Museum of Nature & Science has been able to partner with the BLM and WIPS to build a growing collection of more than 250 species of fossil plants and a similar number of fossil insects. We're still waiting for our first bird and butterfly, but the excitement of that possible find was with the crowd that had assembled around Mike at the base of the Radar Dome hill.

Mike sported Coke-bottle glasses, and a bouquet of hand lenses hung from his neck. He was surrounded by a motley crew of individuals, couples, and families. Some were clearly there for their first experience and had no idea what they were in for. Others were veterans of the Green River and were equipped with spatulas, aprons, crowbars, shovels, brushes, dinosaur T-shirts, flip-up

sunglasses, and other esoteric accoutrements of the well-dressed paleonerd. Mike gave a brief overview of the plan, handed out his 128-page Xeroxed manual, and set his small army loose on the side of the hill. Ray and I walked around the hill to look at the view from the purdiest site. When we got back about 30 minutes later, fossil insects were starting to spill from the outcrop, and the troop was as happy as pigs in lake-bottom mud.

One little girl had found part of a large five-lobed leaf, a real prize. Lucky for her, the leaf was a thing known as *Macginitiea wyomingensis* and, besides being spectacular, was one of the most common fossils on the hill. She had no need to fear that her treasure would be confiscated for science.

Knowing that we would get to see the best of these fossils back in Denver, I pried Ray loose from the outcrop just as he was getting into a good patch of rocks and herded him into the truck. In the light of day, the precarious pitch was visible to Ray, who urged me to watch the road. Of course, these words were wasted, as they're always wasted when a geologist is driving. Telling a geologist to watch flat pavement or gravel when the roadside is composed of gorgeous road cuts is like asking a fashion writer to watch the runway instead of the models. It's just not going to happen. You simply must trust that the geologist at the wheel is used to driving with his eyes off the road.

What a view it was. To the south, our vista was a spectacular shot of a cliff and a winding road that meandered down the forested valley toward Fruita. It was easy to see the contact between the cliff-forming white layers of the Green River Formation and the underlying slumping sandy layers of the Cretaceous Mesaverde. On the other side of the truck, the road was bordered by a steep rock wall where we were given a close-up view of layer after

layer of oil shale. As the road carried us steeply down the hill, we descended lower and lower into the bottom of the old lake. At one point in our descent through this thick phone book of time, we came to a place where the shale appeared to be folded like a shoved carpet. We hopped out to inspect it and realized that we were looking at a layer of stromatolites, slimy mounds of algae that formed along the shores of the great Eocene lake. Their presence here, so far from the mountains, was pretty clear evidence that the depth of the lake fluctuated greatly and that, at times, the whole lake was quite shallow. Pondering the implications of a lake that was as large as Lake Erie but possible to wade across, we headed off to Meeker. "Ah, never a dull day when cruising with the geologically savvy," wisecracked Ray. "Who would ever guess that staring at rocks could be so much fun?"

The wistful
loneliness of the
geologically
inclined

12
AMMONITE SPAWNING GROUNDS

We stopped in Meeker for lunch and were attracted by a little log-cabin county museum. We spent a few minutes looking at the usual assembly of arrowheads, World War medals, and purple glass bottles before stumbling on the sad tale of the Meeker Massacre. In 1879, Ute Indians got seriously annoyed with an Indian agent named Meeker who treated them poorly and plowed up their pony-racing track. They killed him and a few others and kidnapped a young woman named Flora Ellen Price. There was a lovely little photograph of Flora. Ray also admired a small but very bloody oil painting of the massacre that reminded him of his own childhood battlefield paintings. The Meeker Massacre inspired terror in the settlers of Colorado, but it was far more unfortunate for the Northern Utes, who were shipped off to a reservation in Utah.

We continued up the valley of the Yampa River through the coalfields of Craig and Hayden. As the valley tightened and cliffs came down to the road, we left the thick sand layers of the coaly Mesaverde and started to pass by layers of dark gray marine shale. We were going back out to sea and into ammonite country. Just before

Steamboat Springs, the road cuts started sporting big round concretions, and we both knew that only a sledgehammer lay between us and cool fossils.

We challenged Big Blue to climb Rabbit Ears Pass. It's a long hill, and we had to stop once to let the engine cool. The downhill run was easy, and soon we were in Middle Park and near the site of one of our most successful digs. When I moved to Denver in 1991, I heard rumors of giant ammonites in the mountains near the town of Kremmling. When Ray and I started to talk about building the *Cruisin' the Fossil Freeway* exhibit in 1998, I followed up on the rumors. It turned out that the giant ammonites of Kremmling were no secret. They were great big mothers called *Placenticeras*, found in giant Milk Dud–shaped concretions. The average specimen was about two feet in diameter, but some of them were almost three feet wide. Thoughts of ammonites the size of car tires raced through both of our minds.

The site, a nondescript sagebrush slope located a few miles north of Wolford Mountain Reservoir, is littered with giant split concretions with hollow ammonite-shaped cavities where the ammonites have been removed. When

Probing for ammonite nodules in a field near Kremmling.

these depressions are filled with water after a rainstorm, they look like birdbaths. The BLM office in Kremmling had protected the site, but only after collectors had hauled away all the exposed ammonites. When I asked Frank Rupp, the local BLM resource guy, if there were any ammonites left, he told me that there were only birdbaths. The BLM had fenced the site, put up a few interpretive signs, and told people how to find the place if they asked.

So in 1998 I contacted Emmett Evanoff, a Boulder-based Colorado geologist who had done some work in Kremmling in the 1980s. Emmett had a cool idea. He thought that the concretions were in a discrete layer, and if we could trace the layer, we could find more unopened concretions buried beneath the surface. That made a lot of sense to me. Emmett thought we could organize a big group of people to walk across the field in a row, each armed with some sort of thin steel probe, poking the ground to find the shallowly buried concretions. I got the idea that we could use pitchforks to probe the ground more thoroughly. Our early tests showed that it was too hard to plunge all of the tines into the ground, but eventually, we learned to cut off the two middle tines so that only the outer two remained. I organized a band of museum volunteers, Ray and his family came down from Ketchikan, Frank issued us an ammonite digging permit, and we set out to probe for giant ammonites.

The two-tined ammonite probes worked like a charm, and by the end of the first morning, we'd found six buried nodules. I set teams on the project of digging up the massive Milk Duds, and by midafternoon, several were exposed. We still had no idea how we were going to go about cracking open the 300-pound rocks, but that turned out to be almost no problem at all. The ammonite shells are a source of calcium, and the thirsty roots of the prairie plants had already weakened and cracked the nodules. The bad news was that the roots had completely destroyed the original shell material. The good news was that their handiwork made it simple to lift off the top of the nodule and pull out the ammonite. It was fun work and reminiscent of opening a sarcophagus. Soon the bed of Big Blue was full of giant ammonites. The ammonites were so large that four

(left)
Big Blue full of giant ammonites.

(left middle)
Sexual dimorphism of ammonites and humans.

(left bottom)
On our ammonite safari, life imitates art.

A bellyful of ammonite.

of them filled the truck bed. Within a few days, we had probed, dug, and collected a dozen of the giant shells.

Why the ammonites were there in the first place was a question that had been torturing Emmett for a long time. *Placenticeras* ammonites, like humans, are sexually dimorphic, that is to say that the girls are a different size and shape than the boys. In the case of humans, the males are, on average, larger. For ammonites, the opposite is true. Our big nodules were filled with girl ammonites. But where were the guys?

Emmett, Ray, and I spent some time discussing this with squid experts to see if we could extract any clues from modern-day ecology. The nearest living relatives to ammonites are squid, octopuses, cuttlefish, and chambered nautiluses. I visited the Smithsonian Institution, where the squid curator, a crusty Yankee named Clyde Roper, told me of his quest to capture a giant squid and ushered me into the squid collection, a huge room full of pickled squid and cuttlefish carcasses.

We learned that some species of squid spawn simultaneously, just like salmon. Ray had a lot of experience with salmon and was familiar with the concept of how they swim around eating for four years and then all come home to their stream of birth, have a massive orgy, and then die. Streams in Alaska in the fall are filled with the carcasses of spawned-out salmon. Perhaps the giant ammonites near Kremmling were also spawned out. It turned out that some modern squid do get together to spawn but, unlike salmon, the males and females separate after having sex, each gender going its own way. This seemed a plausible explanation for why we primarily found the big female ammonites. We'll never know if this

"ladies' night out" hypothesis is correct, but it was enough to set Ray off on months of frenzied drawing and painting the sex lives of extinct ammonites.

We left Kremmling and drove up the valley to the Colorado River through the drab badlands of the Troublesome Formation, full of fossil camels and small saber-toothed cats, past Hot Sulphur Springs, where a highway department worker found a fine Cretaceous cycad, and into the little town of Granby. We stopped for dinner at the Willow Creek Bar and Grill, a classic joint shaped like a huge wagon wheel. A slouchy waitress wearing a "Bite Me" T-shirt slid into the booth with us and took our orders, then disappeared. We got into a deep hour-long conversation about fossil hunting along the Oregon coast and didn't notice that our food hadn't arrived until the owner showed up and told us that he'd fired the waitress. He asked what we wanted for dinner, and shortly thereafter we enjoyed a couple of very nice, juicy, and free hamburgers. We camped in a forested valley, and the next morning collected some Paleocene leaves before loading the truck and heading west toward Utah.

For all of its fossils, Colorado is a little slim on the classic Paleozoic marine fossils so common in the mid-continent and Appalachia. The exception to this rule is the Pennsylvanian Minturn Formation, which is stuffed with brachiopods, corals, bivalves, crinoids, and even the odd trilobite. Fortunately, we were in it. We stopped by the road near the McCoy cemetery and crawled around finding spirifers, the archetypal brachiopods.

A couple of years ago, I was visited by the mayor and sheriff of Minturn, a little hamlet not far from Vail. Both the mayor and the sheriff were impressive barrel-chested men, and I was curious why they had driven to Denver to see

The Minturn mayor's fossil tree.

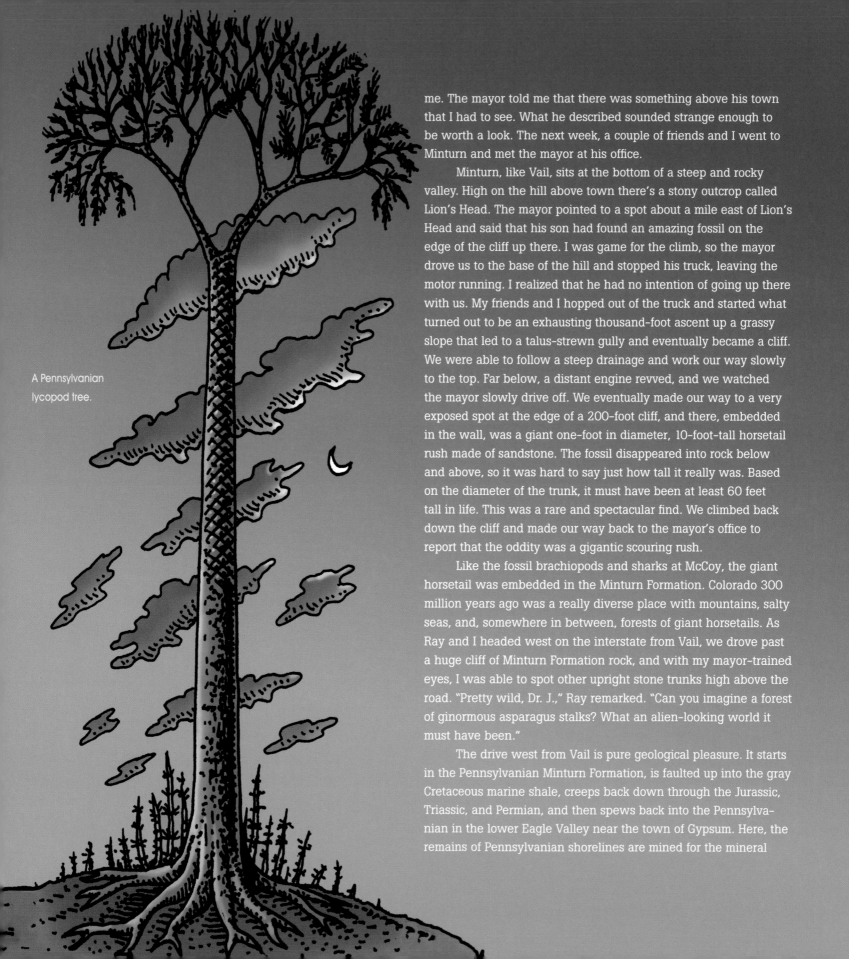

A Pennsylvanian lycopod tree.

me. The mayor told me that there was something above his town that I had to see. What he described sounded strange enough to be worth a look. The next week, a couple of friends and I went to Minturn and met the mayor at his office.

Minturn, like Vail, sits at the bottom of a steep and rocky valley. High on the hill above town there's a stony outcrop called Lion's Head. The mayor pointed to a spot about a mile east of Lion's Head and said that his son had found an amazing fossil on the edge of the cliff up there. I was game for the climb, so the mayor drove us to the base of the hill and stopped his truck, leaving the motor running. I realized that he had no intention of going up there with us. My friends and I hopped out of the truck and started what turned out to be an exhausting thousand-foot ascent up a grassy slope that led to a talus-strewn gully and eventually became a cliff. We were able to follow a steep drainage and work our way slowly to the top. Far below, a distant engine revved, and we watched the mayor slowly drive off. We eventually made our way to a very exposed spot at the edge of a 200-foot cliff, and there, embedded in the wall, was a giant one-foot in diameter, 10-foot-tall horsetail rush made of sandstone. The fossil disappeared into rock below and above, so it was hard to say just how tall it really was. Based on the diameter of the trunk, it must have been at least 60 feet tall in life. This was a rare and spectacular find. We climbed back down the cliff and made our way back to the mayor's office to report that the oddity was a gigantic scouring rush.

Like the fossil brachiopods and sharks at McCoy, the giant horsetail was embedded in the Minturn Formation. Colorado 300 million years ago was a really diverse place with mountains, salty seas, and, somewhere in between, forests of giant horsetails. As Ray and I headed west on the interstate from Vail, we drove past a huge cliff of Minturn Formation rock, and with my mayor-trained eyes, I was able to spot other upright stone trunks high above the road. "Pretty wild, Dr. J.," Ray remarked. "Can you imagine a forest of ginormous asparagus stalks? What an alien-looking world it must have been."

The drive west from Vail is pure geological pleasure. It starts in the Pennsylvanian Minturn Formation, is faulted up into the gray Cretaceous marine shale, creeps back down through the Jurassic, Triassic, and Permian, and then spews back into the Pennsylvanian in the lower Eagle Valley near the town of Gypsum. Here, the remains of Pennsylvanian shorelines are mined for the mineral salts that formed by the evaporation of a 300-million-year-old sea. A factory turns the old shoreline into plasterboard walls.

Near the town of Dotsero, the Colorado River joins the highway and both plunge into Glenwood Canyon. The steep road plummets downhill and into lower and deeper rocks, eventually passing all the way through the Paleozoic to the Precambrian before reaching the town of Glenwood Springs. At Glenwood, a huge fold known as the Grand Hogback brings the Pennsylvanian rocks back to the roadside and this geology creates the town's famous hot springs. A few miles west of town, a roadside outcrop of Cretaceous limestone is studded with meter-wide clams. A little farther west, a ridge of Late Cretaceous sandstone and mudstone contains a seam of coal that, to this day, burns underground and visibly smokes when the ground is snowy.

The bedrock flattens out into the Piceance Creek Basin, and, just past the nondescript town of Silt, the brilliant Roan Plateau rises to the north. The face of the Roan is the flat-lying oil shale of the Green River Formation. In these cliffs north of Rifle, the Green River Formation has yielded its only fossil scorpion, a two-inch beauty named *Uintascorpio meyershawesi*. West of Rifle and below the Green River cliffs, red-and-tan-striped badlands of the DeBeque Formation crop out on the north side of the road. Lens-shaped sand layers in the DeBeque are cross-sections of ancient stream channels, and the passing motorist has the opportunity to casually glance at a perfect cross-section of a 56-million-year-old landscape.

In the 1930s the fossils of the Roan Plateau got their bard in the form of Grand Junction newspaperman and amateur fossil hunter Al Look. Al was a passionate promoter of the prehistory of the region, writing books such as *1,000 Million Years on the Colorado Plateau* and *In My Back Yard*. Look spent a lot of time hiking around the badlands and cliffs, and there was a lot to find. His biggest was in the DeBeque Formation, where he hit a mother lode of Paleocene mammals.

Paleocene mammal fossils are fragmentary pieces of small-bodied animals, reflecting the destructive heritage of the Cretaceous extinction that quite literally killed all of the large land animals on the planet. The survivors ranged in size from raccoon-small to shrew-tiny, and it took much of the 10 million years of the Paleocene for their evolutionary descendants to regain large body size. The DeBeque Formation was deposited late enough in the Paleocene that bear-sized mammals were back in the picture.

Look found pieces of some of these Paleocene big boys and contacted Bryan Patterson at Chicago's Field Museum. Patterson, whose father was famous for killing Uganda's famous man-eating lions in 1898 and writing about it in *The Man-Eaters of Tsavo*, was a self-educated scientist who never graduated from college but ended his career as a professor at Harvard. Together, Patterson and Look recovered nearly complete skeletons of *Barylambda* and *Titanoides* (originally known as *Sparactolambda*). *Barylambda* was the size of a black bear and it had a long otterlike tail. *Titanoides* was a saber-toothed, flat-faced monstrosity. Look chronicled these lost worlds with a whimsical and clever pen reminiscent of Mark Twain. And just like Twain, some of his tall tales were true. Look considered the whole of the Colorado Plateau to be his backyard. It's a fine tradition carried on by some residents of Grand Junction today.

Trees Are Made of Gas

Photosynthesis is not common sense. We all learned it in school, but not many of us really internalized the fact that plants are made of air and water. Sure, plants do contain a trace of mineral matter from the soil, but that usually accounts for less than 5 percent of their weight. The next time you look at a mighty oak, a baseball bat, or a pumpkin, think about the fact that you're looking at what happens when carbon dioxide gas and water are combined with the aid of a little solar energy. When you're looking at a tree, you're looking at gas made solid.

Lycopod trees, the scaly forest giants of the Mississippian, had a very peculiar relationship to photosynthesis. Detailed studies of their anatomy show that, like conifers and broadleaves, they had the ability to transport water through tiny tubes in their trunks. But unlike these familiar trees, they had little ability to transport the products of photosynthesis back to the rest of the plant. As a result, lycopods had to localize their photosynthesis and, thus, they would have been green from the tops of their roots, all the way up their trunks, to the tips of their leaves.

On our original trip, Ray and I passed on the opportunity to drive up the valley of the Roaring Fork River from Glenwood Springs and on to Carbondale, Basalt, Snowmass, and Aspen. At the time, there wasn't a compelling reason to take the drive. The Roaring Fork is a famous trout stream but neither of us are fly fishermen so that was not a draw. I did recall that the Denver Museum had collected an interesting fossil from the valley above the little hamlet of Old Snowmass in 1967. In 1966, a couple kids were hunting with their dad when they came upon a large rib cage protruding out of a bank of dark gray shale. They did the right thing and called the Denver Museum to report their find. The next summer, a team was dispatched to inspect the site, and they realized that the gray shale was deposited by the Western Interior Seaway. The fish was a *Xiphactinus*, the same giant fish that is well know from the chalk beds of Western Kansas.

Xiphactinus are truly impressive beasts, and the largest ones are up to 17 feet long. They have teeth that would make a medium-sized tyrannosaur proud, and they were clearly voracious predators in a sea full of other dangerous animals There have been several finds of whole *Xiphactinus* with the bodies of slightly less large fish in their rib cage. This is the way of some large predatory fish who ambush large prey and then swallow them whole and headfirst. It seems that it was often the case that *Xiphactinus* would eat fish that were just a little too large for their own good and the act of eating the smaller fish

The Snowmastodon shovel army.

Gary Staab uncovers a splendid mastodon mandible.

caused the death of the larger fish. This may be one of the oldest know exampled of the perils of greed, but it surely makes for some amazing fossils.

My mention of a giant Cretaceous seaway fish in an outcrop near a famous trout stream got Ray thinking about fishing the Cretaceous seaway with a dry fly. The Snowmass fish was 13 feet long and bent into the shape of a *U*. It didn't have anything in its stomach so it is hard to say what killed it but it was an amazing fossil and me made me think of fossils whenever I heard the word Snowmass.

That concept got supersized on October 15, 2010, when I received a call from the Colorado Geological Survey that someone has found a mammoth skeleton near Snowmass Village, a mere eight miles from Aspen. I love those kinds of calls, and I dispatched Bryan Small, one of the museum's fossil preparators, to drive up to Snowmass and check out the find. It was a mammoth all right—a pretty complete skeleton of a young female mammoth. It has been discovered when a bulldozer operator named Jesse Steele had driven his D-9 right through the skeleton and ribs had come up over the blade. Snowmass Village had recently acquired the water rights to a small lake and their intention was to drain the water from the lake, excavate a large amount of lakebed sediment, build a dam, and create a deep reservoir that they could use to make snow for the nearby ski mountain, and also have enough water to supply the local hotels and restaurants.

It seemed like a typical salvage paleontology situation, and we made plans to go dig the rest of the skeleton. But while we were preparing ourselves, the Roaring Fork Valley caught a bad case of mammoth fever. The local water and sanitation district started displaying the bones in their office and thousands of people came to see them. Meanwhile, it grew increasingly unclear who owned the skeleton. The contractor thought that he did because he was being paid to carry away the dirt. The landowner thought that he owned the fossils because they were on his property and all he had done was to sell the water rights, not the mammoth rights. The Water and Sanitation District thought they owned the fossil because they had purchased the water rights, and Snowmass Village thought that they owned the fossil because it occurred within the confines of the village. This made Pitkin County think that they owned the fossil because it was in the county. I knew that the skeleton would come to the Denver Museum because that was the place where the state of Colorado stores their fossils, and this was clearly a Colorado fossil.

I headed up to Snowmass on October 27 to straighten out the ownership issues, meet the players in this drama, and inspect the fossil site. When I arrived at the 12-acre drained lake, I saw a large white tent that the construction company had erected over the portion of the mammoth that was still in the ground, and I saw that the bulldozers were back at work moving sediment and loading it into trucks to be hauled away. I was wearing my talk-to-lawyer clothes, but the dirt was calling to me, so I donned a hard hat and walked into the construction site. Almost immediately, I saw a bulldozer turn up a large bone. I was with one of our archaeologists and we started digging with our hands and pulling up bones, logs, and even a piece of an ivory tusk. There was clearly more than one mammoth in this old lake. That changed everything. It wasn't a single mammoth, it was an Ice Age site, and it was at 9,000 feet above sea level.

I drove back to Denver that night and started to plan a big dig. The next morning I got a text showing three construction workers holding three ivory tusks and another image of an engineer holding a mastodon tooth. Now we had evidence of both mammoths and mastodons. I scrambled to send a larger team to Snowmass and decided to go back myself. Between October 28 and November 17, it was one big, glorious fossil scrum. Fossils were everywhere, and we added the giant *Bison latifrons*, Jefferson's Ground sloth *Megalonyx jeffersoni*, and a compete fossil deer skeleton to the species list. We worked like fiends until heavy snow shut down the construction site and the fossil site for the winter.

Winter gave us the time we needed to plan a proper excavation plan and we negotiated a 50-day dig from May 15 to July 1, 2011. We recruited a team of 37 scientists from 19 institutions and a digging army of more than 300 volunteers. When spring finally arrived, we hit the mud running and started finding a steady stream of amazing fossils. On some days we found as many as 300 large bones. Most of them were from mastodon but we also found evidence of horses, camels, and host of smaller mammals, birds, reptiles, and amphibians. It turned out that the site had two distinct levels, a lower level full of mastodon, bison, and sloth and an upper level full of mammoth, deer, and camels.

As the dig wore on, I kept calling for more diggers and I absolutely demanded that Ray fly down from Ketchikan to be part of the find. Ray arrived with artist Gary Staab on June 13 and joined in the festivities. Both Ray and Gary had the joy of discovering large, beautiful Ice Age bones and Ray even wrote and recorded a song about the site.

When the smoke cleared at the end of the 69-day dig, 380 diggers moved more than 8,000 tons of dirt and extracted 5,426 large bones by hand. It was the quickest, largest, and most fun dig in the history of Colorado. And every time I hear the word Snowmass, I think of fossils.

13
THE DINOSAUR DIAMOND

Grand Junction, a peach-growing town located in the Grand Valley at the confluence of the Gunnison and Colorado rivers, is a place where fossil-obsessed citizens have had a century of field days. Mount Garfield, named for the assassinated president, towers high above the northeast side of the town. Capped by thick sandstone layers, its flanks are smooth gray slopes nearly devoid of vegetation. This is the Mancos Shale, mud from the bottom of the Cretaceous Sea. This formation is full of marine fossils and is the westward equivalent of the super-fossily Niobrara chalk beds of Kansas, both part of an 80-million-year-old sea that stretched from western Utah to at least Saint Louis.

"Dino Jim" Jensen pulled out a 50-foot-long mosasaur, *Prognathodon stagmani*, near Cedaredge, and giant clams and ammonites can be found with diligent digging. The Mancos also hosts the exquisite Cretaceous crinoid called *Uintacrinus.* There's debate about whether these gorgeous creatures lived on the seafloor or dangled from floating logs. Fossils like these transform gray shale hills into libraries of lost worlds. Approaching Grand Junction, we saw that the Mancos, which towers above the town and is what underlies it, is now regularly victimized by dirt bikers whose ubiquitous tracks had scarred even the steepest slopes.

The road south from Junction leads to the Black Canyon of the Gunnison and the giant dinosaur site known as Dry Mesa. A local couple, Eddie and Vivian Jones, found the site in 1971 and showed it to Dino Jim. He toiled for more than a decade there, collecting enough bones to fill the basement of the football stadium at Brigham Young University. In addition to the standard Morrison fauna, he found the huge theropod *Torvosaurus* and the giant sauropods *Brachiosaurus* and *Supersaurus.* Jim made himself famous with a picture of his stocky frame dwarfed by a huge shoulder blade.

Grand Junction itself is home to 45,000 enterprising souls who make a living in agriculture and the service industry. In 1900, Chicago-based paleontologist Elmer Riggs wrote letters to people in Grand Junction asking if they'd seen any old bones. A dentist wrote back with affirmation, and on July 4, 1900, Elmer found a giant dinosaur just a few miles outside of town. Then and there, Grand Junction became part of the expanding world of dinosaurs.

The beast from Riggs's Hill, *Brachiosaurus altithorax*, was the same species of long neck that sneezed on the little girl in the first *Jurassic Park* movie. Although Riggs's skeleton was only about 20 percent complete, it was enough to qualify as North America's largest dinosaur skeleton, and, for many years, a cast of this absurdly large beast was mounted in Stanley Hall at the Field Museum.

When the Field Museum acquired *T. rex* Sue, the giant plastic brachiosaur was shipped to United Airlines Concourse B at O'Hare Airport. I was always so impressed at the skeleton when it was at the Field, especially at how much larger it is than the *Diplodocus* we have on display in Denver. Yet somehow, the setting at O'Hare diminished the old boy. His tail pokes out over the security area, his head looks longingly at a Starbucks. I'm not sure that dinosaurs and air travel mix. Then again, since we now know that birds are dinosaurs, maybe they do.

Grand Junction is the largest town on the Colorado Plateau and, as such, is also a center for the mineral exploration that has rocked the region since the Manhattan Project ramped up demand for uranium. Al

Look's 1956 book, *U-Boom*, chronicles the craziness that came down as every Tom, Dick, and Harry grabbed a Geiger counter and became a prospector. All that prospecting led to a lot of incidental fossil finds, and a lot of old prospectors eventually turned into rock and fossil hounds.

The uranium deposits of the plateau were formed as uranium-rich groundwater seeped through bedrock and found places to precipitate. A lot of those places turned out to contain buried dinosaur bones and fossil logs. People were mining dinosaur bones for their uranium content or simply finding radioactive bones as they searched for ore. But uranium was not the only mineral to precipitate in the bones and wood: often fossils were agatized with a beautiful red silica. Pretty soon, prospectors were cutting cabochons out of agatized dinosaur bone or fossil logs.

Today, Grand Junction is full of rock hounds, dinophiles, and fossil nuts. My friend Dick Dayvault, a reclamation specialist contracted by the Department of Energy, collects petrified logs, seeds, cycads, bennettites, and cones on this western edge of Colorado and the adjacent Colorado Plateau in Utah. His friend Frank Daniels, a former district attorney, has become one of the finest collectors and photographers of polished petrified wood. In 2006, Frank and Dick published *Ancient Forests: A Closer Look at Fossil Wood*, a full-color, 456-page exposé of two men's obsession with gorgeous fossils. In Denver, I get e-mails from Grand Junction physicians and math professors who hunt fossil leaves in the nearby Book Cliffs. Either there really is nothing to do in Grand Junction, or the fossils are really that good. Could be a little of both.

Ray and I stopped at the Museum of Western Colorado in Fruita. It used to be in downtown Grand Junction, but when a competing museum run by a company called Dinomation failed, the Museum of Western Colorado took over the Fruita space 10 miles west of Grand Junction. The new operation, christened Dinosaur Journey, has a highway exit and is next door to a McDonald's. It is poised for tourist success.

Dinomation is a California-based company that makes robotic dinosaurs—big rubberized beasts that move a little and roar a lot. During the heyday of *Jurassic Park*,

when Hollywood finally realized that 99.9 percent of all children love dinosaurs, Dinomation provided quasi-accurate beasts that traveled the country appearing in museums. These popular exhibits seemed really thin on science but drew big crowds. When Dinomation closed up shop in Fruita, they left the big rubber models, and the Museum of Western Colorado, itself a legitimate museum with actual research collections, inherited an odd legacy of oversized toys.

Among the first monsters that greeted us as we walked into the museum was a roaring *Utahraptor* in the process of ripping the head off a hapless baby sauropod. It was hard to escape the tawdry thrills of the cheap horror-flick sensation. We chatted with the curator, John Foster, who was working hard to nurture a group of retired volunteers.

Hunger gripped us, so we stepped out of the cool museum into an asphalt-melting 106-degree midday scorcher and headed across the interstate into Fruita town center, a quaint touristy place replete with dinosaur sculptures, dinosaur murals, dinosaur paintings, and dinosaur street names. Thankfully, we found an air-conditioned Mexican restaurant on Main Street. Western Colorado in early July was a little more than Ketchikan-based Troll had bargained for. The restaurant was good and we took advantage of it, eating bowl after bowl of tortilla chips. The second we sauntered back out onto Main Street, we realized we were toast. Acknowledging that full bellies and summer heat are why the siesta was invented, we started rearranging our plans.

My friend Rob Gaston lives near Fruita, and I had never lived up to my many threats to visit him. I rummaged around in the toolbox of Big Blue and found my trusty paleontologist's little black book. We called, he was home, and we headed over, perfectly ready for a slow, lengthy, indoor, preferably air-conditioned conversation. Rob and his wife, Jennifer, greeted us like they'd been planning our visit for months. Rob is a lanky Tennessee boy with the trace of an accent and a casual style. He runs a business selling exquisitely detailed plastic casts of prehistoric animal parts.

Rob is one of those paleo-obsessed natural history buffs who has found a novel way to make his obsession pay the bills. Wandering around his studio was very much like walking around a candy store, or a chocolate store, to be precise. Rob uses a nice hard plastic, makes beautiful molds, and generates replicas that look indistinguishable from the real thing. But because he makes multiples, the little rows of brown raptor claws, *Allosaurus* teeth, and oviraptor skulls really looked like tasty little chocolates from the Rocky Mountain Chocolate Factory.

Rob had his start working for Lin Ottinger, a rock shop owner in Moab, and while working for Lin, Rob discovered a dinosaur bone bed in the Cedar Mountain Formation. Some of the bones belonged to an odd armored ankylosaur that was new to science. In time, Utah paleontologist Jim Kirkland named the new genus *Gastonia* after Rob. The site, known as the Yellow Cat Quarry, was extensively dug by teams from the College of Eastern Utah in Price and eventually

yielded the first good specimens of the giant raptor known as *Utahraptor*. We talked about this as we drank our iced tea. It wasn't a siesta, but Rob's southern style was just the slowdown we needed.

Ray and I pushed off before our visit turned into an unintentional self-invitation to dinner. We didn't make it far. Exit 2 in Utah is known as Rabbit Valley, and it's a mandatory paleostop. To the average cross-country driver, this no-service exit holds no attraction, but to the fossil-wise, it's a treasure land. The bounty begins at the bottom of the off-ramp. We pulled Big Blue into the empty parking lot and climbed out into the still, hot afternoon to take a walk. I love this spot because it's possible to literally stand on a 20-foot-long sauropod neck without knowing it.

Sauropods are not called long necks for nothing. Because their necks are so long, it's easy to ignore the incredible intricacies of each vertebra.

In fact, for long necks to function, each of the component bones had to be an engineering wonder. A single neck vertebra on a garden variety *Camarasaurus* might be

Rob Gaston and the skull of *Gastonia.*

DINOSAURS + AIR TRAVEL - A STRANGE MIX

159

the size of a keg of beer, but it's an intensely elaborate bone composed of a central disk and a whole series of projections of thin bone. Fossils of elaborate things are tricky to interpret because the sediment that buries the bone eventually hardens to rock that breaks in a manner that's irregular with respect to the convoluted bone. Sauropod necks are really bad this way.

I strolled ahead of Troll and sauntered onto the neck. He ambled up after me, mumbling something about the heat before he grumbled, "So, professor, where in the heck is this cool fossil site?" I mentioned that he ought to be more careful when standing on a dinosaur's neck. We spent the next half hour crawling around the enigmatic fossil as the interstate traffic roared below us.

After leaving Rabbit Valley, we drove up-section through the Cedar Mountain Formation and the Dakota Sandstone before ramping onto the broad plain of the Mancos Shale. This really is the land of buried treasure. Over the long years of the uranium boom, prospectors scoured this landscape, and barroom tales of whole dinosaurs and perfect fossil logs abounded.

As we continued west, the north side of the road revealed an endless rampart of stepped cliffs. Known as the Book Cliffs, these sandstone ramparts stretch all the way from Grand Junction to Price, almost a hundred miles. I explained to Troll how the filling and emptying of the great Cretaceous Seaway had paved layers of beach sands on top of the seafloor mud. The sea slowly filled and emptied, the region slowly sank, and each time the sea refilled, mud would be deposited on top of sand. When the sea emptied again, sand would be deposited on top of mud. In this way, hundreds of feet of alternating sand and mud

accumulated. The mud was full of marine fossils, the sand full of beach and swamp fossils. Then, when the sediment turned to stone, the sand hardened more than the mud, so that when erosion carved the Book Cliffs, the sandstone made vertical cliffs while the mud made slopes. The layers of the Book Cliffs each have their own name, but collectively they're known as the Mesaverde Group, a reference to the giant sandstone cliffs near Durango that are famous for their cliff dwellings.

Over the years, I'd explored a number of the canyons that sliced through the cliffs in my search for Cretaceous fossil leaves. Up Thompson Canyon, the remains of old coal mines were telltale signs of ancient vegetation, and there we found fossil leaves from the forests that grew on the shore of the sea. Broadleaf trees, palms, ferns, and conifers related to bald cypress composed the flora. Even though it was 75 million years old, the vegetation would have looked similar, at first glance, to what you'd see today if you boated around the swamps near New Orleans.

The Book Cliffs layers formed at the same time as the rocks at Dinosaur Provincial Park in Alberta, which is the richest dinosaur site in the world, yielding more than 40 different kinds of dinosaurs. This would make you think that the Book Cliffs should be a real dinosaur mecca. It hasn't been, probably because the acid groundwater associated with swamps is a strong chemical deterrent to the fossilization of bone. Dinosaurs were here but were not so commonly fossilized. To be sure, there were some exceptions; geologists have found a few foot bones of a tyrannosaurid, possibly *Albertosaurus*, and a Utah student found the back end of a mummified hadrosaur. The reality is that the dinosaur fossil action to be had was to the south of the interstate, not the north, in what is called the Cedar Mountain Formation.

Long a fairly obscure formation sandwiched between the Morrison and the Dakota, the Cedar Mountain Formation is exposed from the Colorado state line all the way to Price, Utah, and from there to the south along the eastern side of the San Rafael Swell. It's been a tricky formation to understand because its thickness is variable and the fossils found in it suggest that the Cedar Mountain was deposited at several distinct times during the Cretaceous. So where the Hell Creek Formation was deposited in a 1.5-million-year time span at the end of the Cretaceous, the Cedar Mountain may have been laid down in as many as five pulses, each perhaps a million years in duration, but spread over as much as 30 million years.

Since the first gasp of the Cedar Mountain Formation was being deposited during the first years of flowering plant evolution, you would think that it would be a great place to tell the story of the origin of flowering plants. Unfortunately, the chemistry of the formation is poor for plant fossilization: only tree trunks and logs were preserved with any regularity. One of the sad ironies of paleontology is that the rock layer that you'd guess would answer a specific question often doesn't. Some formations were acidic when they were deposited, and the bones were dissolved but the plants weren't. Other formations were just the opposite: leaves and flowers are destroyed, but bones are well preserved. The Cedar Mountain is a formation that's good for bones and logs but bad for leaves and flowers.

The Cedar Mountain Formation appears to have as many as five different dinosaur faunas, each composed of a different group of animals. Not bad for a layer of rock that was largely ignored during the first, second, and third rounds of dinosaur exploration in the American West. In fact, the heyday of Cedar Mountain dinosaur exploration appears to be underway as we speak. Sparked by Gaston's find at Yellow Cat Quarry in 1990, teams from BYU, the College of Eastern Utah in Price, the Museum of Western Colorado (now Dinosaur Journey), and the Denver Museum of Nature & Science have been searching for new sites

Flower Power

The beginning of the Cedar Mountain Formation was in the earliest Cretaceous, between 125 and 95 million years ago. This was a critical time in Earth history, because it was when the first flowering plants appeared. Flowering plants are confusing because their technical name, angiosperms, is poorly known, and their common name places so much emphasis on the flower. A lot of flowering plants have very subtle flowers, so you couldn't be faulted for missing them. For example, all of the broadleaf trees are flowering plants, as are all of the grasses, rushes, palms, and most aquatic plants. Of the eight major groups of living land plants, flowering plants account for more than 80 percent of the nearly 300,000 species. Number two are ferns, with about 20,000 species, and third place goes to the conifers, with a few hundred species. Cycads, gnetales, and lycopods have just over 100 species each; horsetails a few; and ginkgo just one. So flowering plants are by far the most species-rich and, in general, the most common plants on the Earth today. In a general rule of thumb for botanists, if it's not a fern or conifer, it's likely a flowering plant. The fact that flowering plants date back only to the Early Cretaceous means that the Jurassic landscape probably looked a whole lot different from any landscape you can find on Earth today.

KEN'S
THUNDERING
HERD OF
THIGH-HIGH
ANKYLOSAURS

in this formation and quarrying with great energy. I had visited my museum colleague Ken Carpenter and his volunteer crews at their quarries in the Cedar Mountain Formation and quickly learned that I have neither the skill nor the patience to be a bone digger. At one of his sites, Ken has been excavating a fossilized herd of thigh-high *Gastonia*. The skeletons were scattered during burial, and their remains form a continuous layer 75 yards wide.

Ever aware of our paltry budget, Ray and I checked into a markedly substandard hotel in Green River and lowered our body temperatures by sliding into a pool that hadn't been cleaned since the Cretaceous. As the sun finally set and the temperature became reasonable, we emerged back onto land and wandered over to a burger joint appropriately named Ray's. Having followed Ray into greasy spoons in a dozen states in search of the perfect

hamburger, I wondered if he would be tempted to rank this one higher because of its name.

The next morning, we backtracked a bit to visit Moab. The road from Interstate 70 runs downhill and into lower rocks. We passed rapidly down through the Dakota and Cedar Mountain formations. These exposures of the Cedar Mountain are known to produce beautiful petrified logs. It was near here in the early 1980s that Moab poet and rock hound Frank Lemmon and his wife, Leona, had discovered a petrified cycadeoid trunk that wasn't squat and round like other cycadeoids. It was cylindrical, almost two feet in diameter, and eight feet long. The trunk was brutally heavy, but it had already broken into portable sections. The Lemmons hauled the telephone pole of a rock back to Moab, where it caught the attention of BYU paleobotanist Don Tidwell. The quality of petrifaction was good, so he was able to see the anatomy of the plant on the broken surfaces. What he saw baffled him.

Cycadeoids, similar to many living cycads and palms, armed their squat trunks with persistent remains of old leaf bases, and this trunk was no different. Unlike cycads, which produce cones, and palms, which produce flowers, cycadeoids produced flowerlike structures that were nestled in between the persistent leaf bases. There was a lot of debate about whether these structures ever opened and looked like flowers or if they stayed closed and looked more like pods embedded in the surface of the trunk. Some detailed anatomy on the specimens collected in the Black Hills seems to suggest the latter, a finding that ruined a lot of beautiful prehistoric paintings by artists who had chosen to reconstruct the trunks to look like pineapples covered with daisies. The thing that stunned Don about this fossil was that every single leaf base had an associated flowerlike structure. Did the plant grow to full stature before forming its flowers? Were there leaves all over the thing or just at the top? Why weren't the flowers at the bottom of the pole more mature than the ones at the top? Here he had the most perfectly preserved and unusual Early Cretaceous plant ever found, and it was too complicated to figure out.

We continued toward Moab, passing the entrance to Arches National Park, where Edward Abbey penned his revolutionary tome *Desert Solitaire.* The line of RVs queuing at the ticket booth seemed to mock the book. Abbey was always so mad at people for not taking the time to appreciate the landscape. I always wondered how much he appreciated the vanished landscapes that had come before, the ones that had shaped his world.

Moab is now exactly what Abbey feared: a giant outdoor theme park. I'm sure that most of the slickrock mountain bikers, ATV jockeys, and jeepers have no clue about the prehistoric underpinnings of their red-rock playground. There's at least one major exception to this rule. We'd driven to Moab to meet Lin Ottinger, proprietor of Ottinger's Rock Shop, a squat building surrounded by piles of rocks on the north side of town. Rob Gaston had told us enough about the man that we just had to meet him.

A *Utahraptor* lurks behind a grove of *Monanthesia* cycadeoids.

By the time the second the bell on the front door had stopped ringing, we knew we had walked into a real rock shop. Most rock shops these days buy their stock wholesale at the big show in Tucson. For that reason, you can be at a rock shop in Steamboat Springs or Jackson and see nothing but the same fossils from China, Morocco, and Brazil. Like the rest of the American economy, the rock shop business is globalizing. The problem with that is you learn nothing about the local area. Refreshingly, Ottinger's place was a sampling of the local landscape. There were shelves of gorgeous acid-etched horn corals from the Permian Rico Formation, which is exposed just to the north of town, and big agatized dinosaur bones from the Morrison.

Lin Ottinger, Moab's premier rock shop owner.

We overheard a kid whining, "I really want a prehistoric thing," and we approved. "One of our peeps," smiled Ray. After slowly working our way around the room, Ray took advantage of a lull at the cash register and engaged Lin in conversation. A lanky no-nonsense man, Lin had been selling rocks for 68 years. He was born in Casper, Wyoming, in 1927. His family moved around a bit, and Lin was selling rocks and arrowheads in Tennessee during the Depression, when he was seven. A few years later, his family visited Peterson's Rock Garden, an early rock shop near Prineville, Oregon, and Lin started making career plans. In 1939, at the age of 12, he first visited the Denver Museum. Philip Reinheimer had just completed the Hall of Dinosaurs, displaying a *Diplodocus* from Dinosaur National Monument, a *Stegosaurus* from Cañon City, and an *Edmontosaurus* from Montana. The dinosaurs had a big effect on young Lin.

By the 1950s, he was prospecting for uranium on the Colorado Plateau. He rode the boom and rode out the bust, realizing that this was the place for him. He opened his own rock shop and started taking visitors on backcountry jeep trips and rock and fossil safaris. He appeared in a 1962 issue of *National Geographic*. In 1973, he found a dinosaur in the Cedar Mountain Formation that was new to science, and in 1979, Dino Jim Jensen named it *Iguanodon ottingeri*. Lin combines the crusty seen-it-all crankiness of an old-timer with a true curiosity about rock and fossils. His time is passing. Globalized fossils and regulations that prevent the sale of fossils from federal lands have changed the rules for rock shops on the Colorado Plateau. Nonetheless, Ottinger's story is integral to how we have come to see fossils in the American West.

After taking advantage of the presence of a good latte in jetsettin' Moab, we headed back to the interstate. Past Green River we veered north, avoiding the topography and geology of the San Rafael Swell, and rolled into the sleepy town of Price in the late afternoon.

As the fourth corner of the "dinosaur diamond" that includes Vernal, Grand Junction, and Moab, Price seems aware of its past. "Dino" is the high-school mascot, so we felt no need to start our usual mascot campaign, and signs to the College of Eastern Utah Museum are everywhere. The museum is a modern building full of not only fossils, but also some really interesting archaeological exhibits. We'd come to see paleobiologist Don Burge, but he was away, so the registrar, a pleasant fellow named Dwayne Taylor, took the task of showing us around. He told us that in 1988, the museum acquired a spectacular and nearly complete mammoth from Huntington Canyon north of Price. The mammoth was discovered at an elevation of 9,500 feet by an alert backhoe driver. Ray started riffing about enlightened backhoe drivers who have the good sense to stop. "How many times do you hear this story in the ever-expanding western suburban sprawl: 'Backhoe operator finds prehistoric wonder.' Think of how many times they don't stop! I think museums should launch an outreach program geared to construction workers!" Ray was building up steam. Just as he was about to step on the next soapbox, Dwayne interrupted with, "Want to see something really interesting?"

Then Dwayne took us to the museum's off-site storage facility in an old hospital. It seemed appropriate to be storing old broken bodies in an old hospital. You could tell that Burge and his team had been busy in the Cedar Mountain by the room after room of prepared dinosaur bones Dwayne showed us. The museum had worked three main quarries in the Cedar Mountain Formation: the Yellow Cat, which we already knew about based on our conversations with Rob Gaston, the Mussentuchit, and the Price River #2. The Mussentuchit had produced a different fauna that included a group of juvenile ornithopods known

as *Eolambia* as well as an armored dinosaur known as *Cedarpelta*. There was a whole room full of the shiny black bones of baby *Eolambia*.

The Price River #2 is located in the Ruby Ranch Member of the Cedar Mountain, about 20 miles from Price. This quarry had a lower level that gave up parts of five brachiosaurs. Its upper level produced a huge nodosaur, an armored dinosaur nearly 35 feet long.

Besides Gaston's *Gastonia* and Ottinger's *Iguanodon*, the Yellow Cat Quarry had also yielded *Nedcolbertia*, a cute little carnivore named after Ned Colbert, the discoverer of the big *Coelophysis* bed at Ghost Ranch, New Mexico. The big Yellow Cat find for Burge came in 1992 when his team found parts of a huge raptor that would later come to be known as *Utahraptor*. The timing of the discovery could not have been better. That was the summer of *Jurassic Park*, and Steve Spielberg had a problem. Michael Crichton had loved the name *Velociraptor*, a dinosaur first found at the Flaming Cliffs of the Gobi Desert by Roy Chapman Andrews in 1924, and had shaped the book around a pack of those dinosaurs gone bad. No mind that *Velociraptor*, though probably vicious, was smaller than a turkey. Crichton had amplified his ideas from John Ostrom's work on *Deinonychus*, the nasty wolf-sized dinosaur from Montana, so it wasn't too egregious to borrow the wicked-sounding name *Velociraptor* for the larger but less aggressively named *Deinonychus*. The problem started during preparation for filming, when Spielberg decided that even *Deinonychus* wasn't big enough for the sense of threat he wanted to impart. Remember, this was the guy who made *Jaws*. So Spielberg did what any self-respecting filmmaker would do, he invented a larger dinosaur, a double-sized *Deinonychus* that he called *Velociraptor*, or "raptor" for short. Enter Don Burge, the Yellow Cat Quarry, and *Utahraptor*. By the time the movie had opened, giant raptors were a reality rather than a Hollywood exaggeration. *Utahraptor*-mania blazed in the hearts of paleonerds across Utah. Soon T-shirts and books were plastered with its snarling visage.

We asked Dwayne about *Utahraptor*, and he invited us into one of the museum's inner collection rooms where

Utahraptor, the giant raptor from the Cedar Mountain Formation that made Steven Spielberg an honest man.

an old bank vault stood. We stood back as he rolled the tumblers, and within a minute Ray was holding the giant six-inch slashing claw of the original *Utahraptor* in his trembling hands. It really was better than holding one of Rob Gaston's near-perfect reproductions. This was the real thing, the business end of what was very clearly an animal that neither Ray nor I would like to meet in any other form than stone-cold dead. We thanked Dwayne profusely for the opportunity to shake hands with the killer and wandered back into the museum exhibit area.

Before we headed out, we admired a display of dinosaurs from the nearby Cleveland-Lloyd Quarry, a famous Morrison site that we planned to visit later in the trip. Parts of an *Allosaurus*, a *Camarasaurus*, a *Stegosaurus*, and a *Camptosaurus* lay strewn on the exhibit floor, testimony perhaps to the lack of a person with the skills to mount them upright. It was not the last time that we would see this type of lay-'em-flat exhibit style.

Price is coal mining country. The Cretaceous rocks of the Book Cliffs are coal rich, and, to the north and west of Price, underground coal mining has been active since the 1920s. These coal mines are world famous because of their dinosaur tracks. When the coal was a swampy forest, it was apparently quite a busy place, with a variety of dinosaurs—mainly hadrosaurs—marching around in the muddy water, leaving tracks all over the peaty floor of the swamp. Eventually floods brought in mud and sand, which filled in the tracks and buried the peat. Seventy-five million years later, when the peat had changed to coal and the sand and mud was the roof of the coal mine, Price-based coal miners noticed that when they removed the coal, giant three-toed rocks would fall off the ceiling of the mine. This was a bit of a worry, because a miner's hard hat is pretty useless if a 100-pound rock is dropped on top of it. The dinosaur tracks became a safety hazard, and there are unconfirmed rumors of miners who were actually killed by the tumbling tracks, perhaps the only humans ever killed by dinosaurs.

We loaded back into Big Blue and headed up Price Canyon en route to Ogden, Utah. As we passed over Soldier Summit, we encountered deposits of the westernmost edge of the giant Eocene Green River lakes. Sheets of lakeside shale from this area are covered with fossil bird footprints. Big Blue made its own tracks as the Rolling Stones, blasting from the stereo, carried us out of the Colorado Plateau.

Breathing Some Life into Those Old Bones

"Mounting" a dinosaur skeleton is not an easy feat. Heavy, because they're petrified, and often distorted, crushed, or cracked by the process, dinosaur bones are a real challenge to the would-be museum designer. There are museums that dodge this problem by mounting plastic or plaster casts. Some scientists argue against mounting real bone because it makes it less accessible for scientific study. But people's valid desire to see the real thing continues to encourage museums to mount actual skeletons. It's done with supporting steel bars and cables, and a good mount does a thorough job of minimizing the obviousness of the supporting steel while maximizing the vitality of the dinosaur's pose.

Ken Carpenter at the Denver Museum of Nature & Science is the Zen master of dinosaur mounting. For 25 years I've watched him breathe life into tortured fossil bones, creating skeletal mounts whose vitality belies their fossilized reality.

14
THE MORMON TRILOBITE CHOIR

We burst out of the Rockies and into the Basin and Range at Ogden. The huge Salt Lake was hidden from sight, and the swell of the rapidly growing urban sprawl of the Wasatch Front quenched any sense that we were on a drive in the Old West.

Stop number one was the Eccles Dinosaur Park in Ogden. Utah has the dinosaur sickness bad, and this place is one of the symptoms. Like the dinosaur garden in Vernal, Eccles Park has gone to great pains to rebuild the old boys in fiberglass. The number of different dinosaurs that you see when you walk around the park is quite amazing. Director Brooks Britt is a serious dinosaur paleontologist, and he's added a smallish museum to the front of the operation where a few skeletons are lying around on the floor. It was a nice afternoon, so we wandered around the grounds trying to figure out why people pay to walk among plastic dinosaurs. But there were moments of discovery as we rounded a bend and saw a few vaguely lifelike models. Ray posed and roared beside the *Parasaurolophus*, and we experienced an intimidating sense of scale by standing beneath a life-sized *Brachiosaurus*.

We'd come to this part of Utah to see the Gunthers, a fossil-finding family that lives in nearby Brigham City at the foot of the Wellsville Mountains. Three active generations of Gunthers are fossil fanatics, and the fourth generation is now playing in the quarries and is well on its way to joining the family hobby. The patriarch of the tribe is Lloyd Gunther, who, though he's nearly 90, has the handshake of a grizzly bear, the result

of years spent with hammer, crowbar, and shovel in the search of trilobites.

When I first met Lloyd in 1991, he was married to his second wife, Freida, and the two of them, with their combined families, had a sum of 120 children and grandchildren. I asked Freida how she remembered their birthdays, and she said that it didn't really matter because there was one every few days. Lloyd's son Val and Val's son Glade round out the central triad of the bug-digging clan. The Gunthers' forte is trilobite hunting, and no one does it better. For years they headed south to the Cambrian beds west of Delta and prospected thousands of feet of Cambrian shale in search of better bugs. Unlike the masses that returned time and time again to the known trilobite holes in the Wheeler Quadrangle, the Gunthers explored new areas, making fantastic discoveries and passing on their finds on to museums around the country. In 1981, BYU published the Gunthers' *Some Middle Cambrian Fossils of Utah*, a definitive text on the more than 40 species of trilobites found in central Utah.

I met the Gunthers shortly after they made a major discovery a few miles from their house. The Wellsville Mountains are the northern extension of the Wasatch Front, and they rise abruptly and steeply off the flat surface of old Lake Bonneville, the Ice Age lake that made the Great Salt Lake look like a puddle. Knowing that the Spence Shale was Middle Cambrian in age, Lloyd and Val surmised that they might be able to find trilobites there. It was really

Lloyd Gunther, patriarch of the Gunther trilobite tribe.

THE TRILOBITE SUBCULTURE

tough work, because the rocks of the Wellsville Mountains are steeply tilted and the layers are angled into the hill. This is the worst possible configuration for digging. To make matters even worse, the Spence Shale is very hard rock. But the Gunthers are nothing if not persistent, and they managed to find a thin vein of shale that was rich in trilobites. Then, to their amazement, they started finding more than trilobites. They found worms, hyolithids, bizarre arthropods, and scalelike sclerites and realized that they were finding Burgess Shale creatures.

The Burgess Shale is a world-famous Cambrian site from the Cambrian shale of the southern Canadian Rockies. Originally discovered by Charles Doolittle Wolcott in 1911, a single quarry at an elevation of nearly 12,000 feet preserved a whole host of bizarre animals that had their popular debut in Stephen J. Gould's book *Wonderful Life*. The Burgess fauna was thought to be unique, but then a similar assemblage known as the Chenjiang was found in China. Shortly after that, the Gunthers found yet another assemblage a stone's throw from their own backyard. I visited them in 1991 and Val, Lloyd, and Glade took me and a couple of my elderly volunteers to see their site. It was a brutally steep and slippery climb up a talus slope of hard shale, and I was impressed that Lloyd's old limbs seemed immune to fatigue.

I was further amazed when Glade whipped a heavy rock drill out of his pack, fired it up, and started boring a hole into the mountainside. A little while later, Val ran a wire down the hole and poured liquid in behind it. Then he backed away, crouched behind a rock, whispered, "Fire in the hole," and set off a detonation that blew rock and dust high into the sky. This came as a huge and unwanted surprise to my crew, who were looking for trilobites in the talus slope just below the cliff. But for just a moment, it was raining trilobites.

Later on that same trip, I learned how fast an 80-year-old could move. Apparently, in the congenial but competitive world of rock hounds, there's an unspoken fossil-hunting rule that whoever puts his hand on a fossil first is the one who gets to keep it. Lloyd and his son Val had taken me and some museum volunteers to a few of their favorite trilobite holes so that we could collect material for the *Prehistoric Journey* exhibit. In addition to trilobites, I was particularly keen to get a few of the rarer Cambrian fossils, things such as the shrimplike phyllocarids or arms of the predatory *Anomalocaris*. I was talking to Lloyd about this as we stood next to our vehicles, which were parked at the edge of a trilobite hole in the middle of a huge sagebrush flat. As we spoke, we both gazed over at a pile of shaley slabs. At the same instant, we both saw the big phyllocarid on top of the pile. I started to point it out, but I was interrupted by a quiet "whoosh" as the large grandfather lunged past me. Lloyd covered 10 yards in an instant, and the next thing I knew, he was holding the fine fossil in his hand. Never mind that he'd brought the museum to this spot to find this very kind of fossil. This one was his. Lloyd and Val eventually loaded my truck with fossils for our exhibit, so I didn't begrudge him his phyllocarid, but it did teach me a lesson about the stealth and speed of spry octogenarians.

The Gunthers are shale diggers of the first order, and they'll dig anything that's preserved in flat fissile rock. Lloyd had found a Cretaceous fossil leaf site near Henefer, Utah, less than an hour's drive from his house. Whenever he got the fossil-digging bug, which was often, he would head out to Henefer with his hammer and crowbar and bust up the road cut. As a result, he'd acquired a huge collection of these leaves. Knowing that I studied Cretaceous fossils, he offered me the collection. I drove to Brigham City to pick it up and haul it back to Denver.

Now, a few years later, Lloyd had filled his workshop again, generously offering me this new collection, so I was back for another load. Ray and I spent an hour or so with the hand truck loading the back of Blue with almost 60 Coke flats full of fossil leaves.

Coke flats, usually seen next to Coke machines, are the box of choice for fossil hounds. Cheap and easy to find, the flats also hold a reasonable weight of fossils.

After we were finished loading, Val came over and we spent some time looking at the display cases that crowd Lloyd's living room. Although he'd recently given most of his collection to a new museum in Cedar City, Utah, Lloyd had held back a substantial selection of premium bugs. His third wife, DeEsta, joined us and brought out some paintings that she'd been working on. Here was a real Grandma Moses, only her paintings were of family fossil digs rather than family picnics. Glade joined us and chanted an old Pavant Ute Indian saying about trilobites: "Timpe-Konitza-Pachuee." Roughly translated, it means "Little water bug living in a house of stone."

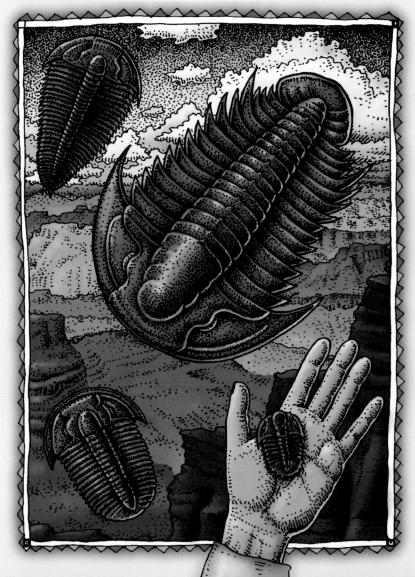

It took a while to disengage from the Gunthers, and when we got back on the road, the truck was substantially heavier with all the fossils we were hauling. We headed south, wishing that there were more family fossil dynasties in our great nation.

We arrived in Salt Lake and headed up the hill to the Utah Museum of Natural History at the University of Utah. This is a classic university museum, with little thought given to how the public might visit it. Testimony to this are

the five parking spaces available to the public. Luckily, we scored one. Our goal was to visit the chief curator, Scott Sampson, to hear about his work in the North Horn and Kaiparowits formations. Scott is a tall, handsome guy who looks far too normal to be a dinosaur paleontologist. We sat down in his office and discussed dinosaurs. Ray was particularly impressed because Scott was part of the team that named a new species of dinosaur from Madagascar after Mark Knopfler, the guitarist for Dire Straits. Scott said that the same Dire Straits tape was playing repeatedly in the quarry when they found the skeleton, and there really wasn't much choice in the matter.

Scott's team had recently recovered a partial *Tyrannosaurus rex* skeleton from the North Horn Formation at North Horn Mountain in central Utah. This was exciting news, because the original excavations at North Horn by Smithsonian paleontologist Charles Gilmore in the 1930s had yielded the ceratopsian dinosaur *Torosaurus*, the titanosaurid sauropod *Alamosaurus*, as well as a bunch of three-foot-long lizards and a big tortoise. The discovery of a *T. rex* here meant that the legendary carnosaur and sauropods had coexisted. This was music to Ray's ears. "Wow, so artists can now draw *T. rex* and long necks in the same image without fearing ridicule. Cool!"

The possible existence of the K–Pg boundary layer in central Utah was of interest to me, because it would represent the westernmost exposure of the boundary in North America. If there were any geographic variation to the way plants and animals responded to the killer asteroid, this would be a good place to look for it. We got instructions to the site before heading off with Scott and his grad student Mark Loewen, a rounded bear of a guy, to a nice dark cave of a student bar where we drank a few tall beers and inhaled some pizza.

After the first round of beers, Scott's student opened up a bit and told us about his efforts to find dinosaurs in the huge expanse of Grand Staircase National Monument near Escalante, Utah. I had driven through the area in the 1980s, and I remembered how a bend in the road had revealed an awesome vista of blue badlands steeper and more ominous than anything South Dakota has to offer.

The Blues are composed of the Late Cretaceous Kaiparowits Formation. A Berkeley turtle paleontologist named Howard Hutchison told me he was prospecting the Blues and came across a chunk of dinosaur bone that had fallen out of the outcrop. When he picked it up, he realized that he was holding a nearly complete and almost perfect *Parasaurolophus* skull in his hand. *Parasaurolophus* is a duckbilled dinosaur with an exceptionally long skull projection, or crest, that is an extension of its nasal chamber. Duckbill specialist David Weishampel has speculated that *Parasaurolophus* was the Pavarotti of the dinosaurian world and was able to use the crest as a kind of trombone or saxophone. By adjusting its breath, the animal could probably have had quite the melodic range.

This area used to be one of Utah's big backcountry secrets, but when the Clinton administration declared it a national monument at the end of 1996, a spotlight was thrown on this beautiful and remote place. The administrative switch to national monument status led to the imposition of a whole new set of regulations and a mandate to catalog the fossil resources of the region. This was

Tom Williamson of the New Mexico Museum of Natural History and Science in Albuquerque sports a skull of *Parasaurolophus* that he collected in the badlands of the San Juan Basin.

173

both good and bad news for Scott. His team was funded to search for and excavate dinosaurs, but they weren't allowed to use vehicles to do it. They had spectacular luck finding a host of new dinosaurs, including an unusual ceratopsian, an albertosaur, and a *Parasaurolophus*, but they had to hand-carry all of their heavy field gear deep into the wilderness.

Scott's dinosaurs were beginning to suggest that the Grand Staircase was the southern equivalent of Dinosaur Provincial Park in Alberta. By collecting plants and animals up and down the spine of the Rocky Mountains, he stood a chance to resurrect an ancient geography and see how it varied. After days of hot roads and plastic dinosaurs, it was nice to be in a museum where science was driving the questions. It was also nice to be in a cool basement drinking cold beer.

In the years that have passed since our trip, a lot has happened to the Utah Museum, the Kaiparowits, and Scott Sampson. In 2011, the University of Utah opened a spectacular new museum south of campus. Not only did it have lots of parking, but it also did a great job of interpreting Utah's natural history, past and present. The Kaiparowits Plateau has become a mecca for dinosaur paleontologists and also the focus of a nasty political battle that saw its boundaries shrunken during one presidential administration and expanded in the next. Scott Sampson is now the director of the California Academy of Sciences.

Ray had been corresponding with Jim Madsen (*Allosaurus jimmadseni*, a little dinosaur found at Dinosaur National Monument, was named after him), who operates a dinosaur casting business out of a warehouse in the Salt Lake industrial area. Lots of museums need fake dinosaur skeletons, and Jim is one busy man. Even when you find a pretty good skeleton, you usually need some spare parts to round out the mount. When we were mounting India Wood's *Allosaurus* for the *Prehistoric Journey* exhibit, for example, we had to "order out" for all sorts of extra ribs, arm bones, and feet to make the poor thing whole. Ray wanted to scope out Jim's operation, and I had never been there, so we called ahead and told them we'd be coming around. Jim wasn't in, but his son was overseeing a crew

of half a dozen guys diligently working away on a variety of molds. The shop was strangely still because everyone was listening to a book-on-tape about World War II. We could tell that we'd arrived at some point during an invasion, because Jim's son was reluctant to leave the main room and seemed really distracted. He eventually warmed up, put the tape on pause, and showed us around the warehouse, where parts of all manner of prehistoric creatures lay stocked in multiples on large shelves. Their product list boasts 50 different mammals and 60 different dinosaurs. Ray started angling for a freebie, and, amazingly, a few months after he returned home to Ketchikan, he received a box containing a lovely plastic Eocene killer pig skull of his own.

We continued south along the Wasatch Front and stopped near American Fork to visit a new museum, The North American Museum of Ancient Life. It's an absolutely gigantic box of a building located next to Interstate 15 in the middle of a big, themed, mall-like commercial development. Coincidentally, we had arrived for the museum's first anniversary, and one of its founders, Cliff Miles, was happy to tour us around.

Cliff is another art-paleo guy. He started his career with an art undergrad degree and found himself working as a fossil preparator at BYU. Later, he and two of his buddies split off from BYU and opened their own fossil quarrying and preparation business. They had reopened the famous Bone Cabin Quarry near Como Bluff, Wyoming, and were making a decent living selling dinosaurs to Japanese museums. I'd seen some of their fossils when I visited the Gunma Museum of Natural History in Japan and had met Cliff when he and his partners generously donated a new Jurassic ankylosaur to the Denver Museum. The animal

ALLOSAURUS CLAW

was new to science, and Ken Carpenter named it *Gargoylesaurus parkpinorum* after Jeff Park and Tyler Pinegar, the two guys who found the skeleton.

I was eager to see Cliff's new museum for several reasons. The place was huge and reputed to be full of cool fossils, including the country's first *Supersaurus* mount. They'd built the museum with amazing speed, from bare ground to open doors in less than two years. And the thing that really fascinated me was that it was a for-profit museum without an explicit collections or research mandate. In a way, it was the logical extension of the commercial fossil trade.

Cliff was a lively and proud tour guide. We could see that he'd really thrown his heart into the effort and had made a credible showing. The side-by-side mounts of *Brachiosaurus* and *Supersaurus* were truly awesome because of their insane size. We were also particularly taken by a sweet uintathere skeleton from the BYU collections. The skeleton took on larger stature when we learned that the guy who had found it later fell to his death off a cliff. Fortunately, you don't hear that too often in paleontology.

The museum also had moments of tragicomedy, like the reconstructed massive head of the shark *Carcharocles megalodon* that was blasted through a wall as if in an attempt to eat the visitors, or the skeleton of a mammoth being attacked by a bunch of human skeletons. Cliff told us that one was inspired by the film *Jason and the Argonauts,* a favorite of his from childhood, which featured a gigantic battle of skeletons.

We made a fleeting stop in Provo to see the BYU fossil collection and decided it was time to get back to the real outcrops. Our immediate goal was North Horn Mountain, and I insisted on taking the scenic route through Fairview, which put us on the very dusty and slow Skyline Road that creeps along a gorgeous ridgeline. We got nowhere fast and ended up pitching camp in an alpine meadow. The next morning, we squeezed Big Blue down a jeep trail and eventually found ourselves on the north shore of Joes Valley Reservoir. All we had to do to get to North Horn was drive to the south end of the lake and a few miles up the valley. As we approached, smoke began to cloud the sky, and by the time we reached the south end, we found our path blocked by a forest service truck. He pointed to the southeast where flames could be clearly seen licking up the slopes of North Horn Mountain. A helicopter was dipping water out of the lake and working the fire, and the road was truly closed. There wouldn't be a North Horn *Tyrannosaurus* for us that day.

It ended up being a blessing in disguise. We were feeling more like a big breakfast than a big hike, and there were other theropods in our immediate future anyway. A steep drive down the canyon from Joes Valley Reservoir abruptly spilled us from the mountain forests onto the desert of central Utah. Soon we were happily raiding our cooler for a midmorning picnic at the Cleveland town park.

Sated, we headed for the Cleveland-Lloyd Quarry, one of the most enigmatic of all the Morrison deposits. The road to the quarry is a long gravel drive from Cleveland across fairly featureless sage flats that eventually give way to a small escarpment. At the foot of the escarpment

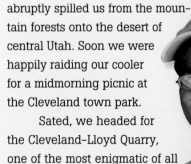

Celina and Marina Suarez, stegosaur-studying sisters from San Antonio.

is a trio of small and markedly unimpressive buildings. We parked and went in to the one-room visitor's center. A laconic ranger chewing on a soggy unlit cigar and two tiny Hispanic women greeted us. Mike Leschin, the ranger, was in his fifth season at this lonely post, and Celina and Marina Suarez are twins from San Antonio, Texas, who'd somehow managed to catch the dinosaur bug. In addition to greatly increasing the ratio of women and twins in Utah paleontology, they were studying stegosaurs with Scott Sampson. The trio had been forewarned by Scott that we were coming, so they were clearly waiting to entertain us. Away we went on a private tour of a very odd quarry.

Cleveland-Lloyd doesn't look like much until you lay eyes on the quarry map, a spectacular diagram that shows all 12,000 bones that have been dug out of the quarry. Most of the bones have been removed, and all that remain are a few bones and a big conundrum.

Carnivorous dinosaurs are far more common at this site than herbivorous dinosaurs. The quarry has produced parts of 44 *Allosaurus*, 2 *Marshosaurus*, 2 *Stokesosaurus*, and 1 *Ceratosaurus*. That's a lot of meat-eaters when you consider that the herbivores in the quarry are represented by only 5 camptosaurs, 4 stegosaurs, 3 camarasaurs, 1 *Barosaurus*, and 1 ankylosaur. That's 49 carnivores to 14 herbivores, hardly a fair fight. Mike didn't stand a chance either, as the tour belonged entirely to Celina and Marina, who told us the tale of carnivore death at high speed, each talking over the other in a steady stream of dinosaureze. I'll admit that the story was a lot more interesting in stereo, but at the end of the day, the murder mystery had an unsatisfactory resolution.

The superabundance of carnivores suggests that there was some sort of trap that lured them to their death. The classic example of this is La Brea Tar Pits, where Ice Age carnivores got caught in sticky tar. Other carnivores couldn't resist the opportunity to dine on their mired colleagues, and they too got trapped. The result is a deadly chain letter of treacherous dinner invitations. The tar at La Brea has yielded far more saber-toothed cats and dire wolves than bison, horses, or sloths, and it really looks like the place was a classic predator trap.

The problem with Cleveland-Lloyd is that there's no obvious trap: no tar, no natural trap cave, just a bunch of bodies buried in a three-foot-thick layer of mudstone. This was not a mystery that Ray and I were about to solve that day and, if you can believe it, even Ray was reaching the dinosaur saturation point. We bid adieu to Mike and the charming Suarez sisters and turned once again to the road. It was time for trilobites.

For the last 40 years, there's been one go-to place for trilobites in North America: Delta. I'd blown through Delta once back in the early '80s and had remembered seeing trilobite safari signs. We arrived in town with the thermometer hovering around 105 and immediately sought refuge in a café, making the problem worse by eating a big greasy lunch. The Gunthers had told us that a local banker owned a necklace made of trilobites that had been found in a local archaeological site in Delta, but it was a Saturday, and the banks were closed. We looked through the yellow pages and found a few places that advertised trilobite trips, but nobody was home. I was starting to feel like I should have been more prepared when I remembered Robert Harris. His number was in the white pages, and when he answered the phone, I said, "Are you the King of Trilobites?" He responded that indeed, he was, and that he was going to be down at his shop on Main Street in an hour or so if we felt like dropping by. We killed the hour lying in the shade at the city park and then slowly strolled over to his store.

The King of Trilobites holds court in what must have once been a pharmacy. Behind the plate glass window are a pair of giant plywood cutouts of trilobites wearing crowns. The cutouts refer to *Elrathia kingii*, the little black trilobite that's the coin of the realm of commercial paleontology. Described by one of the earliest American paleontologists, Fielding Meek, *Elrathia kingii* was noted for its unusual mode of preservation. Most trilobites preserved in shale are preserved as impressions. These *Elrathia* were different, having suffered a mineral enhancement that perfectly captured and preserved their upper surface while thickening the body itself. The result is a perfect little shiny black button of a fossil. They weather out of the shale and lie on the surface like so many shiny black scarablike bugs. Collecting them is like picking up coins. These little guys are everywhere, gracing many a bolo tie. There was a Civilian Conservation Corps camp out in the Wheeler Quadrangle in the 1930s, and with lots of workers wandering around the desolate landscape picking up bugs and taking them home, the trilobites began to get a reputation outside of western Utah. They were a rock shop standard from Fresno to Fargo when I was a kid.

Harris got into the business of mining and selling bugs the same year I was born. He'd been working for Fawcett Ant Farms, the company that supplied ants for ant farms. They used straws to collect ants by the thousands, and they also collected fossils. Robert learned the trade from Fawcett and then went big-time. By the 1970s the state of Utah was leasing land in the Wheeler Quadrangle so that people could mine trilobites. Harris estimates that something like half a million trilobites are mined in this tiny patch of Utah each year.

Visiting Harris seemed like going to the heart of commercial paleontology. The man was a brick. Late middle aged, solid around the middle, wearing a white polyester shirt and gray trousers, he could have passed for a small-town banker. He was in the back of his large shop pricing stones and wasn't very welcoming as Ray and I picked our way around the shop looking at baskets of rocks. Ray, whose Soho Coho Gallery in Ketchikan sells the occasional polished stone and fossil, started going into

shopkeeper mode when he found out Harris was willing to sell wholesale. Soon Ray was picking out polished stone hearts made from septarian nodules and Harris started to warm up.

Here was a man who'd been there and done that. Having mined and marketed bugs for almost 40 years, he was probably responsible for getting more than 20 million trilobites into private hands. I asked him if he liked trilobites. "Not really," he replied. "Just a way to make a buck. I'm not even really into rocks. Ten-year-old girls are into rocks." Then he showed us a framed picture of a nice trilobite with a stylish tapering pygidium (the technical term for the back half of a trilobite). The name, *Alokistocare harrisii* (by the vagaries of scientific naming, the bug is now known as *Altiocculus harrisii*), suggested that some scholar had once admired Harris enough to name a new species after him. The scientist was Dick Robeson, a local boy from

Robert Harris, the King of the Trilobites.

Fillmore, Utah, who parlayed an interest in trilobites into a career as a paleontologist at the University of Kansas. I asked Harris what he thought of paleontologists. "Academics are all a bunch of damn liars," he replied. I tried to remember how I had introduced myself. Meanwhile, Ray was taping up flat after flat of polished hearts and handing over his credit card.

Harris, an active commercial paleontologist, was voicing a sentiment commonly heard at the Tucson and Denver shows, where sellers of fossils meet each year to make their big wholesale deals. Academics see fossils as data integral to the understanding of life's history, while dealers see fossils as part of making payroll. Pure academics and pure commercialists don't see eye to eye over a pile of fossils. As a fossil-loving and fossil-buying kid-turned-museum-director, I see value in both sides and am always trying to broker a middle ground. It wasn't clear whether Harris didn't really understand that I was an academic or whether he sensed that my interest in him was genuine, but he began to warm up to me as well.

We asked Harris if he had any recommendations about seeing the famous trilobite digs. He said we should just drive out to his claim in the Wheeler Quadrangle, and it was just fine with him if we dug a few bugs. "I've got a little man named Jimmy out there, tell him I sent ya." Ray shot Harris's portrait in front of the *Elrathia kingii* plywood plaque and we bid him farewell. The King of Trilobites was a man who didn't seem to enjoy the life of a fossil digger. Ray, laden down with 400 bucks' worth of stone hearts, sighed, "Bugs are nothing but business to that guy."

In the truck, Ray fiddled with his pencil as he pored over the Gunthers' trilobite book. "For some people, like me, trilobites are inherently beautiful objects of mystery, grace, and beauty, elegant symbols of deep time," he sighed. "After all, these are critters from half a *billion* years ago. They're so damn

Buck-a-bug
Jimmy Corbett.

pretty!" I had to agree. Trilobites are the pocketable poster children of prehistory. Thousands of species of these little guys attest to a healthy residence on Earth from 545 million years ago until 252 million years ago, a run of 293 million years. The human family, at 7 million years and counting, should be so lucky.

It hadn't cooled a bit, and I stopped at the store for a big bag of cherries and some ice. As we headed out of town, we noticed that the local high-school mascot was the jackrabbit. Ah, the mighty rabbits of Delta. It seemed like the Delta Trilobite Kings might win a few more games, and we added yet another school to our list that needed to go paleo with its mascot. Ray asked if I thought Harris followed high-school football.

The road out to the trilobite hole was long, straight, and dusty, and it took us about an hour to get there. Not a soul nor a vehicle was around as we pulled up the draw to a building-sized heap of shale next to a hole that was about 50 feet deep. We parked the truck and walked down the ramp into the trilobite quarry. There was a beach umbrella at the bottom of the hole, and hunkered beneath the umbrella, chipping away with a hammer, was a small, shirtless, bronzed man. He glanced up as we approached and returned to his work without saying a word. There was no radio playing, no sound at all, just this guy, his umbrella, and a cooler.

Ray approached and said that Harris had sent us. Jimmy didn't say anything. We awkwardly started poking around the shale. Every minute or so, Jimmy would find a trilobite and toss it into a five-gallon plastic bucket. We started splitting shale, but with no luck. Then Jimmy said, "Here's a big one," prying up a dinner plate–sized slab with a three-inch trilobite. Ray flipped. Jimmy said, "You just have to let your eyes adjust and you'll start to find 'em." We started finding 'em.

Jimmy was a wiry piece of lean jerky who probably didn't weigh more than 125. He'd been a blaster for road crews, the guy they hang off the side of cliffs to set the charges. Now he liked working for Harris, getting paid for what he found, about a buck a bug, sometimes more, depending on the size. Harris was a good man who cut a fair deal and left Jimmy alone. On a good day, he could put a hundred bugs in his bucket.

Buck-a-bug Jimmy eventually started talking to us, and the insanely hot day started to cool. It was nice popping bugs and eating cherries. Pretty soon we each had about a dozen bugs and Ray made a deal with Jimmy for the three-incher. We waited until the sun started to set before we left, and Jimmy was still digging when we drove away.

In our push back to Denver, we wound our way through the San Juan Mountains and took time only to dig fossil leaves in Creede. This little 19th-century silver-mining town sits right in the bottom of a huge volcanic caldera, a great collapsed ring that formed a lake 25 million years ago. The lake did as lakes do, and the caldera-filling sediments also trapped fossils. Creede is an unusual fossil site because the caldera lake formed when the San Juan Mountains were already at significant elevation. Here was the rarer sort of depositional setting, a little spot of D-World high in the Miocene E-World mountains. The Creede fossils reflect their elevation, and we found needles and cones of pines and spruces. They looked just like the plants that were growing at the fossil site. Ray and I liked splitting the beautifully thin paper shale, but good fossils were few and far between. Denver beckoned and we headed back to town to clean the truck and prepare for the last leg of the trip.

The angry bugs of Zion

15
ROCKY MOUNTAIN RAIN FORESTS

When I moved to Denver in 1991, I was aware that the Denver Basin was known to geologists as a place that was rich in fossil plants. I was a young paleobotanist, and this was good news. Within weeks of arriving, I was collecting fossil palm fronds at the construction site for the new airport, and this was only one of hundreds of construction sites around the booming city. The rock directly beneath the shallow prairie soil of Denver and Colorado Springs ranges in age from 75 million years to 34 million years, and those rocks were deposited as sediments as the Western Interior seaway began to drain off the continent and the Rocky Mountain Front Range literally grew out of the ground. Seventy million years ago, most of Colorado was at or below sea level. Today the eastern plains are 3,000 to 5,000 feet above sea level and the Rocky Mountains themselves are as tall as 14,000 feet above sea level. This signature topography was created as the sediments of the Denver Basin were being deposited, and those sediments tell the stories of all the ecosystems that came and went as the seaway disappeared and the mountain range grew.

The recession of the 1980s was over, people were building things, and every hole in the ground was a potential fossil site. I had studied the old geology reports from the 1870s to the 1960s that described the fossil plants that had been found in the area around Denver and I had a decent idea of what had already been discovered. I started haunting construction sites and road cuts, and almost everyone one of them held fossils. I began to realize that it was possible to tell the origin story of Denver with fossil leaves.

Ray was profoundly aware that, at the end of the day, I was all about fossil leaves. The next leg of the journey was going to contain a lot of fossil plants.

Thirty minutes south of Denver we hurtled past the town of Castle Rock and the most surprising fossil plant site in North America. In 1994, Steve Wallace, a paleontologist employed by the Colorado Department of Transportation, found some fossil leaves in a railroad embankment on the east side of interstate 25 a few hundred yards north of the town of Castle Rock. I knew Steve well, and he frequently brought me fossils from roadcuts that I would identify for him. This time was different.

He brought me a box of very large fossil leaves, and every single one was a species that I had never seen before. Not only

A near-perfect fossil leaf from the Castle Rock rain forest complete with its drip tip.

that, but the leaves had the distinct features of leaves that are found in wet, tropical rain forests today. I was deeply intrigued.

Steve took me to the site, a road cut on the busiest road in Colorado, and dug into the side of the hill. Leaves began to pour out and in a matter of days we had collected dozens of species. The leaves were fantastic and huge, some of them more than 24 inches long and 18 inches wide. More than half of them had long skinny tips known as drip tips. This sort of leaf tip is found in tropical rain forests where the annual rainfall is more than 100 inches. Today, Castle Rock has an annual rainfall of about 11 inches.

By the end of the third day, we realized that we were digging in an intact layer of fossilized leaf litter, and we started to recognize the remains of upright tree trunks. Near the end of the fifth day, we started finding cycad leaves. As the abundance of cycad leaves increased, I realized that we'd uncovered the edge of a whole cycad plant. Ever the media hound, I called the museum, which alerted Denver's television channels.

It must have been a slow news day, because all three channels sent cameras to the site. We cracked open the final slab "live at five." The split was no disappointment: as we lifted the manhole cover–sized slab of rock, it revealed a spray of 25 three-foot-long cycad fronds. From my live-feed earphones, I could hear the baffled anchors trying to decide why a fossil cycad was newsworthy. I must admit that I understood their confusion. Cycads, after all, are tropical and subtropical plants known mainly to specialists and usually seen only in the rare-

The drill rig that took us 2,256 feet from the grass of the Colorado prairie to the seafloor mud of a 69-million-year-old sea.

plant collections at botanic gardens. This plant was so rare that it was beyond their imagination. We eventually excavated the fronds, the trunk, the whole root system, and even a sweet little cycad seedling. When the dust had cleared, we'd collected the most complete fossil cycad plant ever found.

The 63.8 million year old Castle Rock rain forest is now recognized as the oldest and best-preserved fossil rain forest in the world. Its occurrence in Colorado raises interesting questions about the origin and antiquity of tropical rain forests and drives home the point that the fossil record is nowhere close to giving up all of its secrets.

In order to figure out the age of the Castle Rock rain forest site, I collaborated with Bob Raynolds, a lanky, side-burned, goat- and yak-owning petroleum geologist who is curious about more than just finding oil. Denver is full of people who have and are making their living by understanding Rocky Mountain geology. The layers of rock in the Denver Basin and other Rocky Mountain basins are full of oil and gas that accumulated as the result of the burial and decay of ancient organic matter. Being able to see the third dimension of the Earth allows people like Bob to find oil and natural gas. He was curious about what the fossil rain forest said about ancient climates and how it related to the formation of the Front Range. It seemed to us that the layered rocks in the Denver Basin probably held some secrets about the history of the region. So, one day, Bob walked into my office and said, "Let's drill a well." It seemed like a good idea to drill at the tiny town of Kiowa, in the center of the Denver Basin. Our goal was to retrieve a continuous core sample to test our ideas.

With funding from the National Science Foundation and the state of Colorado, we drilled the hole in the spring of 1999. For a simple idea, execution was a real pain. We drilled round the clock for seven weeks, pulling up a 2.5-inch-diameter tube filled with precious cylinders of rock. Five feet at a time we worked our way back in time. The first 65 feet were sand and gravel from the last few thousand years. Then the core penetrated bedrock that was something like 54 million years old. At a depth of 340 feet, the rock was 64 million years old. At 990 feet, we crossed the K-Pg boundary layer and found evidence of the asteroid that killed the dinosaurs. At a depth of 1,680 feet, we found the lowest occurrence of ground-up granite. Below this level we were drilling into rock that had been deposited at the surface before the uplift of the Rocky Mountains. At 1,800 feet, we were pulling up coal from ancient coastal swamps. At 2,000 feet, we were drilling through a buried beach. Around 2,200 feet, we were pulling up cylinders of gray mudstone that had formed at the bottom of a salty sea. Some of these cores cracked open

to show fossil ammonites, extinct marine animals that were 69 million years old. We stopped drilling at 2,256 feet, confident that we had extracted an adequate sample to record the retreat of the Cretaceous Seaway, the uplift of the Rocky Mountains, the extinction of the dinosaurs, the formation of tropical rain forests along the flanks of the mountains, and a second pulse of mountain building.

We're still doing science on the core and using it as a method to date the fossils in the Denver Basin.

Sixty miles south of Denver, Colorado Springs has one of the most impressive geologic settings of any prairie town in the world. Built at the extreme western edge of the Great Plains, the city looks straight up at towering Pikes Peak, which is only 20 miles away but more than 8,000 feet higher than downtown. Most of the citizens of the Springs look west and see rocks but forget that the city is built on them as well.

I've been excavating fossils from Colorado Springs's city parks since I arrived in Denver. In 1993, a guy was digging a barbecue pit in his backyard when he found

Troll with urban ammonites on his mind.

Michele Reynolds, Beth Ellis, and Regan Dunn demonstrate the diameter of an eight-foot tree stump preserved as a ring of carbon in one of the Castle Rock fossil rain forest quarries.

183

an unusual fossil in the Pierre Shale. He described what he found over the phone and it occurred to me what it was. I hopped in the car and drove down to see him that day. He'd unburied a complete yard-long, pearly shelled, giant baculite, one of the straight ammonites. *Baculites* are common, but complete ones are vanishingly rare. He donated the fossil, which went right onto permanent display in *Prehistoric Journey*.

The south half of the Springs is built on the Pierre Shale. Ammonites and sharks' teeth are common there. I like to call them "urban fossils." The north half of town is built on rocks that were shed as sediments when the Rockies were starting to uplift at the end of the Cretaceous. Dinosaur bones show up in construction sites and in the parks on the north end. The K–Pg boundary itself traces a northwest–southeast line through the middle of town. The parks out to the east contain a menagerie of animals that survived the great extinction.

The area that really caught my attention was a couple of square miles of gullies and badlands called Corral Bluffs that is located less than ten miles east of Colorado Springs. The area had good leaf fossils both below and above the K–Pg boundary and I even found a Cretaceous fossil feather.

In 1998 I collected a nearly complete alligator skull from Corral Bluffs. The site kept nagging at me, and when the Denver Museum hired Tyler Lyson as a curator in 2013, I suggested that he have a closer look at the bluffs.

That turned out to be a really good idea. Tyler is an amazing prospector, and he saw what no one else had seen. Fossil skulls from the first one million years of the Paleocene are incredible rare and only a dozen were known worldwide. Tyler realized that some crumbly white rocks at Corral Bluffs were concretions that contained complete mammal skulls. Once he had cracked this code, he was able to find dozens of early Paleocene mammals, turtles, and alligators at Corral Bluffs. His diligence has turned Corral Bluffs into one of the best places in the world to look at what happened immediately after the dinosaurs went extinct.

The west side of town is where the famous Garden of the Gods park displays some of the most breathtaking geology in the world. This beautiful site and additional outcrops near the base of Pikes Peak in Manitou Springs expose an entire sequence of rocks from the Cambrian Sawatch Sandstone through the Cretaceous Pierre Shale. In the park itself the layers of rock are tilted up and are perfectly vertical.

I was eager to keep moving, so Ray was forced to glimpse the giant vertical planks of red sandstone through the truck window rather than wander amongst them. On the west side of Manitou Springs, we briefly stopped to admire an outcrop of the Sawatch Sandstone that lies directly on the 1-billion-year-old Pikes Peak Granite. I made Ray get out of the truck and put his finger on the contact so he could span 500 million years with just a few millimeters of skin.

A fossil skull of this rare and bizarre *stylinodontine taeniodont* was found in a roadcut near the Castle Rock rainforest.

"The Big Stump," the most impressive fossil *Sequoia* trunk at Florissant, was nearly sawed up and hauled off to the 1893 World's Fair in Chicago.

The Florissant Fossil Beds National Monument is located on the west side of Pikes Peak at an elevation of about 8,400 feet and about an hour's drive from the Springs. This is sacred ground for American pale-ontology. The site was discovered in the 1860s, and scientists working for Ferdinand Vandiveer Hayden's survey made it there in 1873. What Hayden's men found at Florissant was a fossil forest beyond compare. Giant fossil logs, some more than 10 feet in diameter, blocked the Florissant Valley to such a degree that it was hard to negotiate a wagon through them. The site also revealed an amazing fossil lake full of insects, fish, leaves, and birds. The paper shale was so fine that even butterflies, moths, and caterpillars were preserved. Some of the moths and butterflies still retained the pattern of the markings of their wings.

Many scientists and museums have benefited from this rich site. Samuel Scudder, a prolific paleoentomologist from Cambridge, Massachusetts, described hundreds of different species of fossil insects from Florissant. In 1877, Henry Fairfield Osborn, William Berryman Scott, and Frank Speir passed through Florissant on their famous Princeton College paleontological expedition, collecting fossils that can be seen to this day in the Peabody Museum in New Haven, Connecticut. In 1915, the Colorado Museum of Natural History bought a piece of property near Florissant, and the first fossils added to the Denver Museum collec-tions were from that site.

By the 1940s, most of the obvious wood had been carted away, and the petrified forest was turned into a tourist trap that was divided into two warring concessions, the Pike Petrified Forest and the Colorado Petrified Forest. It was an ugly competition as each owner vied for tourist dollars. Eventually the rivalry grew violent, and one of the owners caught a bullet in the gut. Only the Pike Petrified Forest survived.

In the 1950s, a big black sedan pulled into the parking lot of the petrified forest, and the driver got out and introduced himself as Walt Disney. He wandered around for a while before leaving, but Mrs. Disney didn't even get out of the car. A few months later, Mrs. Disney wrote to say that Walt was obsessed with the place and would it be possible to buy a stump. A deal was struck and to this day, you can see a chunk of Florissant at Frontierland in Disneyland.

By the 1960s, Pike Petrified Forest had failed and the land was scheduled to be subdivided. Three formidable women, Estella Leopold, daughter of the envi-ronmentalist Aldo Leopold; Vim Wright; and Betty Willard worked with a Long Island lawyer to craft a campaign to save the land as a federal park. In 1969 the federal government purchased the property and established the national monument.

When I first moved to Denver, the monument was doing a great job of being a fossil preserve but wasn't excavating its fossil resources. In 1994, the monument's first paleobotanist was hired and things started to get better for fossils.

The place was deserted when Ray and I pulled into the parking lot. We poked our heads into a trailer, and at the end of the hall we found Herb Meyer. His paleobotanist friends call him "Herbaceous Mire." Herb is a tall, soft-spo-ken Oregonian who specializes in the fossil plants of the 34-million-year-old Eocene-Oligocene boundary. Since the Florissant lake bed is 34.1-million years old, Herb is in fossil nirvana. He ushered us out into the petrified forest. I'd been there many times, so I knew that Ray was about to be shocked.

The fossil trunks at Florissant are the remains of giant sequoia trees, and they are huge. We walked a loop trail. Here and there we came upon massive stone trunks that utterly dwarfed the adjacent ponderosa pines. The monument has tried to preserve the trunks by building

Fossil cranefly and moth from the Clare family quarry at Florissant.

structures over them, but the buildings have the effect of diminishing the trunks. I was happy when we reached a trunk known as the "Big Stump" standing unsheltered at the base of a little hill. This trunk is every bit of a dozen feet in diameter at its base and 9 or 10 feet tall. There was an attempt to transport this trunk to Chicago's Columbian Exhibition in 1893, but its sheer size defeated the effort. We could still see the rusty saw blade sticking out of the fossilized wood. Herb estimated that the trunk alone weighs more than 60 tons.

Sequoias are restricted to California and southern Oregon today, but they clearly covered the Colorado Rockies 34 million years ago. Apparently, this forest had been killed and buried by a catastrophic mudflow from a volcano off to the west.

There are no active quarries on the monument, but there is a small commercial operation nearby where anyone is allowed to dig for a small fee. Herb joined us as we drove over to see the owners of the little U-pick fossil operation. We were met by Toni Clare, her daughter Nancy, and their horse. The Clares are a fifth-generation Colorado mountain family who have owned the quarry property for as long as anyone can remember. They've opened a small garage-sized scar in the hill that exposes a 15-foot thickness of horizontally layered paper shale. During the summer, they charge people for the opportunity to split shale with razor blades and butter knives to find fossil leaves and insects. But the pieces of paper shale are sort of like lottery tickets: the quarry has produced some

astoundingly rare fossils, including at least three birds and a number of moths, as well as tens of thousands of less-valuable specimens.

Nancy and Toni were mixed in their feelings about their quarry. They asked me again, as they have on many occasions, if the museum would like to come down and do a big dig. About the summer diggers, Toni said, "They fill a box full of shale and I just cringe." They understood that some of the fossils from their little quarry had big-time significance. Even though vacationing families and schoolkids had a great time at the quarry, they both knew the quarry held treasures that belonged in museums.

Nancy had found a complete bird, and we talked about what it would take for my museum to acquire it. She wasn't sure if she wanted to sell it. We talked about how one would go about getting a fair appraisal for such a one-of-kind fossil.

We walked down to a nearby garage, where she pulled out a glass-fronted box and showed us the rare prize. The horse peered over Herb's shoulder and Toni shooed it away with an old Christmas tree.

The fossil was awesome: a 10-inch-long bird complete with fossilized feathers. We admired it for half an hour before we realized that we were hours late for our next stop. We thanked the Clares, said our good-bye to Herb, and drove out of the muddy driveway pondering the worth of a 34-million-year-old squished bird carcass. "I guess it depends if it is a new species or not," Ray said. "How could it not be?" I responded. "There are only a few birds from this formation, and the nearest other formation with complete birds is 15 million years older." I'm not telling how much it cost, but I will say that the Denver Museum now owns this fantastic fossil.

Our real reason for coming to Woodland Park was to visit Mike Triebold's new museum, the Rocky Mountain Dinosaur Resource Center, which had just opened. Ray and I didn't have any trouble finding the place. It's a giant new building on the south side of Main Street in the middle of town. Bill Stein, the museum's number-two man, greeted us as we walked in the door and into one of the largest and most dinosauriferous gift shops I've ever seen. Bill, a lean,

Nancy and Toni Clare's lovely bird.

compact dude, was sporting the company uniform: a pair of camo pants and a long-sleeved tan shirt with the company logo. Bill clearly took pride in the fact that a new museum stood where two years before there had been a parking lot.

The place was a trove of Hell Creek and Judith River dinosaurs from South Dakota and Montana, as well as fishes, mosasaurs, and plesiosaurs from the chalk beds of Kansas. A little video-viewing area had a complete library of classic dinosaur movies: *The Valley of the Gwangi*, *When Dinosaurs Ruled the Earth*, *The Land That Time Forgot*, *One Million B.C.*, *The Lost World*, and more. We were both inclined to sit down and watch *Gwangi*, but Bill was just getting warmed up. He took us into the collection room and started pulling drawers. Bill is a nicotine guy, and he smoked as he pulled fossil after fossil for our viewing pleasure. Pretty soon the little collection room was like a pool hall without the beer. He showed us a big skeleton from Montana that had been named after his kid. So there we were, standing in a smoky room looking at a dinosaur named Sir William. The collection was great, but the secondhand smoke eventually drove us into the prep lab, a big room that buzzed with employees making plastic casts of dinosaurs for unknown destinations. We knew we were running late, so we headed south.

Two hours later, and two hours late, we pulled into Cañon City, Colorado's penitentiary town. There's an old and true tale that in the early days of the Colorado Territory, Denver chose to be the capital, and Cañon City and Boulder were vying for which would get the state prison and which would get the state university. Cañon City got first choice and chose the prison. Cañon City has the long mall- and superstore-strewn entry road that characterizes so many western towns of moderate size. As we coasted past the Wal-Mart, we were stunned to see a life-sized, garishly painted orange and pink steel *Stegosaurus*. Despite its first impression, this was a truly elegant animal. It was, as it turns out, a proud artifact of Cañon City's twin claims to fame: prisoners and dinosaurs.

Two complete *Stegosaurus* skeletons from Cañon City are held by the Denver Museum of Nature & Science. The first was found by Fred Kessler, a high-school teacher in 1937 and the second by a Denver Museum preparator named Bryan Small in 1992. The *Stegosaurus* is Colorado's official state fossil, a designation forced on the state legislature in 1982 by a fierce cadre of fourth-graders who waged a two-year campaign that, in the end, could not be denied. The steel *Stegosaurus* was created at the State of Colorado Fremont Correctional Facility with 4,500 hours of donated inmate-welding time. We returned to pay homage to this doubly significant animal later that evening, but for the time being, we were expected at the Dino Depot downtown.

Dinosaurs were discovered in the Morrison Formation in Garden Park, a valley a few miles north of Cañon City, in 1877, just a few months after the Arthur Lakes discoveries in Morrison. Both Marsh and Cope sent men to Cañon City, one of the first battlefields of the bone wars. Marsh's man, Marshall Felch, worked a quarry that would bear both of their names. The Marsh-Felch Quarry yielded the first *Ceratosaurus*, the first *Brachiosaurus* skull, the nesting *Allosaurus*, and the *Stegosaurus* now on display at the Smithsonian. Cope's teams worked farther up the valley. Near a place named Cope's Nipple, they found a single sauropod vertebra that was gigantic, stretching nearly seven feet from base of the centrum to top of the spine. The same bone on *Diplodocus* was only three feet along the same dimension. Cope named it *Amphicoelias fragillimus*. Later comparisons show this bone to possibly represent the largest dinosaur to ever walk the Earth. It's an unfortunate fluke of history that Cope found only the single bone and that the specimen was lost after his death. Subsequent efforts to locate more parts of this giant have failed.

Much of Garden Park is on BLM land, and the Dino Depot is private-public partnership to highlight and protect the fossil resources of the region. The drive from the Dinosaur Resource Center had been a quick two hours, but in that time we'd swung the full arc of opinions about who should be able to collect and own vertebrate fossils.

The Depot was manned by a pair of fossil-smitten women. Donna Engard was an exhibit designer at the Cranbrook Institute in Michigan. Pat Monaco volunteered there, and the two became friends. When Pat's husband died in 1986, the pair moved to Cañon City to care for Pat's elderly parents and found themselves in dinosaur country. The local BLM office needed help with the fossil resource, and the pair rose to the task. When the Denver Museum started its Paleontology Certification Program in 1990, Donna and Pat were in the first class and were the first to graduate.

It was nearly dark and more than two hours past closing when we pulled into the parking lot. I'd phoned ahead, so they were still there, waiting for us with cannons loaded. Donna is the more level-headed of the pair, but even she was a bit snippy from hunger. Still, Donna and Pat are huge Troll fans, so they had a hard time being too mad about our late arrival. They were ready to close the visitor center and head for dinner, but we begged them to turn the lights back on and let us look around. I hadn't been there in nearly 10 years, and a lot had changed. They had mounted a lovely cast of Bryan Small's *Stegosaurus* on the wall, and beneath it, they displayed a massive fossil tree from the Morrison. Plants are rare in the formation, and this was a real beauty. Pat has a wacky side, and she was soon modeling a cow pelvis on her head as she showed us the education collection. Clearly it was time to get some food.

We drove down a back street to an elaborate restaurant complex called Merlino's Belvidere and wandered into a maze of underground rooms full of chatting diners, bustling waiters, and lounge singers. It was a wholly unexpected expedition into the dinner clubs of the 1950s. Pat added to the illusion by telling us that this was where she and her husband had been married 36 years before. We had a magnificent meal of homemade spaghetti and cavatelli.

The next day was Darwin's birthday. In celebration, we were scheduled to meet Pat and Donna at seven for a tour of the local geology. We set out to negotiate Skyline Drive. The West has its share of white-knuckle roads, and I'd heard that this was a good one. As we approached the turnoff, I spotted a billboard that showed a car full of kids and the phrase "Scream 'til Daddy Stops." Clearly, Cañon City residents were my kind of people.

Skyline Drive is a road that was built for kicks just because the city had surplus inmate labor. It was completed in the 1930s and has been scaring the pee out of flatlanders ever since. A cold wind was whistling as we drove up onto the knife-blade ridge. The road is paved, but that's about all you can say for it. Just a few years ago, Donna and Pat realized that the Dakota Sandstone that forms the ridge is full of dinosaur footprints. Unlike the tracks at Dinosaur Ridge in Denver, these are positive tracks that project off the bottom of the sandstone layers. A group of locals chipped away some of the rock below and revealed a nice series of tracks. We got out of the vehicles and wandered back along the ridgeline to the spot where the tracks were exposed. Pat pointed out a set of four that looked like they were made by a squatting dinosaur. So far Pat and Donna have identified iguanodon, theropod, and ankylosaur tracks.

We got back in the trucks and gingerly drove the ridgeline. Troll, as usual, was uncomfortable driving along the edge of a cliff. Skyline is a one-way drive, thank God, and the exit switches down the east side of the ridge, climbing from the Dakota Sandstone into the Cretaceous marine rocks along the slope. We stopped again and walked up to a ridge of sandstone and limestone and right there, no more than eight feet from the road, I spotted two big beautiful ammonites on the sandstone outcrop. Even Pat and Donna were surprised to see such nice roadside fossils within a stone's throw of Cañon City.

16
THREE HANDSHAKES FROM DARWIN

We drove west from Cañon City, stopping in Salida for a coffee drink called Hell's Revenge before turning south toward the San Luis Valley and New Mexico. At Hooper we passed signs for Gator Farm, a place where they take advantage of natural hot springs to grow alligators. By the time we hit Antonito, Jack Dempsey's hometown, the sky had darkened. It began to rain in Chama, and by Tierra Amarilla the rain had given way to a Seattle overcast. By Cebolla, pink cliffs had risen on both sides of the road. Eventually we pulled up to the understated entrance of Ghost Ranch, a place named for a hanging that occurred there once upon a time.

Ghost Ranch is a place where paleontology and art have brushed shoulders. An adventurous correspondent of Edward Cope explored this valley in 1881 and sent back a smattering of fossil bones, including the bones of a small theropod that Cope named *Coelophysis*. Cope himself passed briefly through the area in 1874 on his way to collect fossil mammals near a town called Cuba in the middle of the San Juan Basin. In 1927, Charles Camp from Berkeley led field trips to Ghost Ranch, where they collected skulls of the big crocodile-like phytosaurs from the late Triassic red beds of the Chinle Formation. But the big find came in 1947, when American Museum paleontologist Ned Colbert dropped by Ghost Ranch for a few days on his way out to do fieldwork at Petrified Forest National Monument. It had taken him some effort to obtain the permits to work in the monument, so he meant not to tarry long at Ghost Ranch. That changed on his third day, when Colbert stumbled upon the find of his life, a bone bed consisting of many skeletons of the coyote-sized *Coelophysis*. This was a breathtaking find, because the little meat-eaters were whole, were many, and were Late Triassic.

Colbert realized that his life had changed, cancelled his plans for the monument, and spent the next several seasons at Ghost Ranch excavating his find. It was an embarrassment of scientific riches, so much so that it wasn't until 1989 that, at the age of 84, he finally published his definitive description of the little dinosaur. *Coelophysis* is now a well-known little beast that populates scenes of the Triassic. In 1995, Colbert published a

sweet swan song of a book entitled *The Little Dinosaurs of Ghost Ranch* in which he tells the story of how *Coelophysis* changed his life. In 1998, *Coelophysis* became the second dinosaur in space when a skull was taken to the MIR Space Station (a Montana *Maiasaura* made it in the 1980s).

There is also an artistic connection to Ghost Ranch. Georgia O'Keeffe was already a famous artist when she first visited New Mexico in 1929. She saw Ghost Ranch in 1934 and was taken by the stark and brilliant landscape and old bones that she found in the arroyos. In 1940 she purchased a small house from the ranch's owners. Her distinctive paintings and famous guests, including D. H. Lawrence, Ansel Adams, Charles and Anne Lindbergh, and Joni Mitchell, brought notoriety to the obscure ranch.

We'd driven to Ghost Ranch neither for *Coelophysis* nor O'Keeffe. We came to see a little animal whose name could not be said. Scientists name new species by the time-honored practice of designating a type specimen and publishing a detailed and illustrated description in a scientific journal. A name is not considered valid until the journal is published. This system has worked well for

more than 300 years, and it's the basis for the science of taxonomy. But problems can arise when species names are given to poorly preserved or partial fossils. In some of those cases, there may not be enough biological information to merit a distinct species, and later scientists have the option to come along and "sink" the name.

This became an issue for Ray and me when Bryan Small from the Denver Museum showed us a photograph of a gorgeous and enigmatic little skeleton that had been collected from the Ghost Ranch Quarry in the early 1990s. The animal looked like a cross between an armadillo and an alligator. Bryan, who was working on the description of the animal with the Ghost Ranch paleontologist, Alex Downs, was cagey about the affinities of the animal, but he already had a name for it, a name that wouldn't be valid until the description was published. So Bryan told us the name and then told us we couldn't repeat it. This only increased our desire to see the actual fossil located at Ghost Ranch.

We turned into Ghost Ranch and drove up the muddy road. As we crested a hill, a sweeping vista of New Mexico cliffs came into view: muddy red hills of the Chinle Formation overlain by massive sandstone cliffs of the Entrada capped by the gray conglomerate of the Todolito.

Desmatosuchus, an armored herbivorous aetosaur.

Vancleavea campi

A SERIOUS PLACERIAS

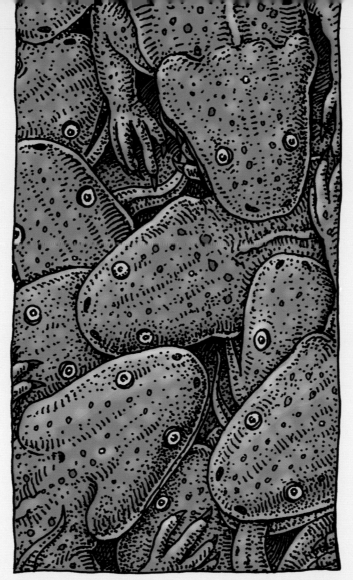

Trees covered the higher slopes of the Morrison Formation, and even farther away, a cliff of Dakota Sandstone shaped the horizon.

We'd called ahead to make sure that Alex Downs knew we were coming, and he greeted us at the door. Before us stood a sprawling bear of a man, his broad sloping frame wrapped in a pair of dueling striped long-sleeved dress shirts, one over the other. His head was hidden under a baseball cap, he wore big, round glasses, and he sported a tangle of alternately light and dark gray scraggily hair that merged into an unkempt beard and mustache. His swollen eyes were watering, and he was wheezing and sneezing. Alex said it was fortunate that we'd called ahead, because he felt like he was going to die, and, were it not for our arrival, he would have gone to the local health clinic. He had clearly made a bad decision on our behalf. We simultaneously warmed to him and gingerly backed away.

He'd been briefed about our obsession with the unnamed animal and ushered us directly to a display case that contained a photograph of the skeleton, a plasticine model of what the animal looked like, and a label that listed the unmentionable name. He said, "Please don't use this name, since it's not published yet." Ray,

rightfully, was confused that we couldn't use it when it was posted in a public place. Alex told him that a poorly preserved skeleton was found in the 1960s in Petrified National Forest and had been named *Vancleavea*, but the specimen was too poorly preserved to be a name holder. Their new specimen was probably the same thing as *Vancleavea*, but it was hard to tell. Once they described the creature in a published article, we were free to use the name. Until then, he requested that we call it an "undescribed archosaur." Still confused, we agreed.

The little paleontology museum at Ghost Ranch is a single room chock-full of the lore of Ghost Ranch and its fossils. There were photos of the early Berkeley expeditions, pieces of skeletons, an ugly diorama of what the place looked like 225 million years ago, and, in the center of the room, a dining room table–sized block of red mudstone completely and utterly covered with bones and skeletons. Apparently, the block had preceded the building, since it was far bigger than any of the doors. The block had been brought down from the *Coelophysis* quarry and was the centerpiece of the room. In the summer, Alex can be found chipping away at it. It was on this block, inside the museum, that he had uncovered the "undescribed archosaur." Ray, ever the fish lover, was delighted to learn the *Coelophysis* quarry was also full of the bones of coelacanths and gars.

Ray became fascinated with another plasticine reconstruction of a long-legged terrestrial crocodile called

A mess of Triassic metoposaurs. These six-foot-long amphibians have been found in large groups.

193

Hesperosuchus. Only a few feet long, the thing reminded us both of the horrible little Gollum in the *Lord of the Rings*. Alex confirmed that Phil Burcheff, the sculptor, leaned toward the Dungeons and Dragons end of things. I asked Alex about Ned Colbert, who had died in 2001 at the ripe Triassic age of 96. He told us that Ned had been around long enough to have worked with Henry Fairfield Osborn, who had met Charles Darwin. By knowing Colbert, Alex was only three handshakes away from Darwin himself. We were about to shake Alex's hand to join this queue of famous paleontologists when Alex exploded into a frenzy of sneezes and hacking coughs. Braving a host of viruses, we both shook hands with Alex.

Finally, Alex ushered us into another building to see the "undescribed archosaur." We entered a cluttered corner room filled with boxes. In the center of the room was a rolling table, and on the table was a tire-sized block of rock nestled in a plaster and burlap jacket. Alex opened a wooden drawer, pulled out a small box, and unwrapped a piece that he then fitted onto the big slab. It was the nearly perfect skull of a most unusual animal. The flattened toothy skull fit neatly into Alex's paw, and he snugged it up against the scaly skeleton. The body was doubled back on itself but was maybe three feet long if extended. There was enough there to imagine the whole animal, and it looked

like nothing I had ever seen in my life: the body was covered with diamond-shaped scales and the legs were quite short relative to the body. It looked like a swimmer, which made some sense in the context of all the fish in the quarry. Suddenly, fossil skull in hand, Alex let loose with another sneezing attack, and I didn't move quickly enough to avoid being sprayed. It was time to leave.

The sun was breaking through the clouds, and the wet outcrops of red Chinle Formation screamed with saturated color as we headed toward Abiquiu. The moon was rising, and a brilliant sun from the west lit up a bank of white clouds above a group of small adobe buildings. I remarked to Troll that the scene reminded me of that famous Ansel Adams photograph, *Moon over … something*. I couldn't remember the name, and neither could Troll. We spent most of the rest of the drive trying to scare up the name from our rusty art history memories. A few days later, Ray suddenly said, "Hernanadez, it's *Moon over Hernandez*." I looked at a map and realized, to my amazement, that we'd been in Hernandez when we saw the reminiscent scene.

Near Taos, we finally turned Big Blue back toward Denver. The sun set and it grew dark as coal. Bruce Springsteen was singing, "I am the nothing man." It was reassuring when, a few hours later, we finally saw the

Folsom point

lights of Raton, a town named with the Spanish word for "rat." I'd last been in Raton with my friend Chuck Pillmore, the USGS geologist who capped his career by locating the K-Pg boundary in the Raton Basin. When Walter Alvarez found the iridium layer in marine limestone near the Italian mountain town of Gubbio in 1980 and proposed the hypothesis that an asteroid had wiped out the dinosaurs, it was Chuck who knew where to look for the same layer in sediments deposited on land. His work with Carl Orth and Bob Tschudy, published in 1981, was the first resounding support for the asteroid hypothesis. After he retired from the survey in Denver, Chuck bought a home high on Raton Pass, not far from the K-Pg boundary. He was an endless source of guidance for people who wanted to see the thin but deadly layer.

In 1992, I took Denver-based dinosaur-track specialist Martin Lockley down to Raton Pass to search for dinosaur tracks below and above the K-Pg boundary. His premise, and it was a good one, was that there should be more dinosaur tracks than dinosaur skeletons. Watching some of my dieting friends rack up 10,000 strides a day on their pedometers, I had to agree. In addition, dinosaur tracks were sure traces of a living dinosaur, while dinosaur bones could have come from a long-dead dinosaur. This all became important as the dispute over the time of the dinosaur extinction grew and some scientists argued that dinosaurs had survived the K-Pg extinction.

There's a small but vocal band of scientists who claimed that dinosaurs became extinct just before the asteroid hit. Another bunch argue that they survived for a while after the impact. I was of the opinion that they disappeared as a direct result of the deadly impact. Martin and I went to sites where Chuck had located the thin boundary layer and found abundant dinosaur tracks below the K-Pg boundary but only alligator, salamander, and bird tracks above the boundary. The track record seems to show the same pattern that we were seeing with the fossil leaves and bones in North Dakota and Montana.

On one of these trips, Martin located a particularly nice *Triceratops* track hanging off the bottom side of a

giant rock that projected from a road cut above the highway. He crawled under the projecting rock and started to scrape away the underlying sediment to better expose the track. I was standing above him watching the top of the rock. As he dug, I noticed a thin but widening crack had appeared in the dirt above the rock. I realized that Martin was undermining the rock that he was lying beneath. I screamed at him to move and, thankfully, he did, because a moment after he rolled out of the hole, the 600-pound rock smashed down on the spot where he'd been lying only seconds before. The rock tumbled down the hill, breaking into smaller chunks as it went. A 100-pound piece complete with the intact *Triceratops* track skidded to a stop within three feet of Big Blue's tailgate, where we easily loaded it up. I'd saved Martin from being killed, and he was appropriately grateful at the averted irony of a dinosaur tracker being crushed by a dinosaur track.

The next morning, Ray and I woke early and sat in the coffee shop at the Oasis. It was a classic old roadside joint with a steady flow of Raton cops and New Mexico state troopers tanking up on their morning coffee. It was the last morning of our road trip, and neither of us were that eager to hit the highway. We had two more famous fossil sites and a long drive between us and Denver, and there was a dinner party waiting for us at the other end. We finished our coffee and headed east to the site of the Denver Museum's greatest find.

Founded in 1900, the Denver Museum of Nature & Science has long been a large regional museum with a bit of a chip on its shoulder. Jesse Dade Figgins, an artist, paleontologist, zoologist, and archaeologist, left New York to become the museum's director in 1910. We were headed to the town of Folsom to see the site of Jesse's biggest success. In the 1920s, the dogma of the archaeological elite stated that humans had not arrived in North America until 3,000 years ago. There was a dearth of archaeological sites to challenge the dogma. Figgins found spear points associated with Ice Age bison bones in Texas but made the mistake of not collecting them in context. As a result, his find had been dismissed as the uneducated screwup of a hick amateur. Figgins had a sense that he was right, and he knew that what he needed was a site where bona fide artifacts were in direct association with the bones of Ice Age animals. It took nearly 20 years, a curious cowboy, and a Raton banker to deliver Jesse's dream.

Driving east from Raton, we passed Capulin Peak, a near perfect 60,000-year-old volcanic cone sitting on the plains of northern New Mexico, and wound down a road that seemed too narrow for two-way traffic. A few minutes later, we pulled into Folsom and back into the 19th century. A summer of driving through little western towns had not prepared us for the perfect little ghost town that we encountered. It was still early and there was no sign of life. When I told Ray we were going to Folsom, he naturally thought of Johnny Cash. This wasn't California, nor was it 1968, but it was clear that time had been draggin' here as well. We got out of the truck and wandered around, wondering when there'd last been signs of life in the town. We quickly found the Folsom Museum, located in a classic old western storefront. A card in the window gave a number to call and said the museum would be opened on request. It was a nice idea, but there didn't seem to be any phones around.

Then I noticed a little post office. I walked in and met a man about my age named Alfred Newkirk. Alfred, wearing a plaid shirt, rodeo buckle, and baseball cap, looked like he should be driving a tractor rather than sorting letters. He told me that in a town of 57, working for the postal service was the best job in town. His parents had owned the general store across the street, which closed in 1987. The nearest doctor was in Des Moines, New Mexico, 45 miles to the east. Of course he knew about the Denver Museum and their big find in 1927, clearly the biggest thing that ever happened in Folsom. He told me that the museum in town was run by two sisters, and he was sure one of them would be happy to meet us. He made a phone call, and about 20 minutes later, Kay Thompson met us in front of the museum. Kay, a short woman in her 70s, wore a patterned fleece coat and the face of a pioneer's daughter.

Kay opened the door and let us into a perfect time capsule. The museum had opened in 1967 in the old mercantile. It contained kitchen gear, bison skulls, saddles,

old rifles, barbed wire, arrowheads, rattlesnake rattles, old photographs, magazines, and newspapers. It was as though anything that ever happened in this little town had been saved for posterity. Ray found an old *Newsweek* with "The Sick World of Son of Sam" blazoned across the front. I found a photograph of Francis Folsom, a wasp-waisted beauty for whom the town was named. She was born during the Civil War and died in 1947. There was a shocking photograph of the hanging of Black Jack Ketchum, a luckless crook who tried to hold up a train near Folsom in 1899. On the east wall hung a sign that remembered "Those hardy pioneers who traveled the Santa Fe Trail."

The museum was great and Kay couldn't have been more helpful, but we'd come to trace the story of a cowboy named George McJunkin, and we were in luck. George was born in Texas in 1856, the son of an enslaved man who saved enough money from blacksmithing to buy his own freedom. George wandered west and learned how to handle horses and cattle. By the turn of the century, he'd settled near Folsom, was managing the Crowfoot cattle ranch, and had a reputation for being well read and curious. It also seems that he was a rock hound. One day in 1908, he found a freshly cut arroyo with bones sticking out of the side of the hill. He mentioned this find to Carl Schwachheim, a blacksmith, and Fred Howarth, a banker, in Raton and got them thinking about it as well. Howarth had recently found a mammoth tusk and was interested in ancient stuff.

George died in 1922, and it wasn't until later that year that Schwachheim and Howarth actually got around to confirming the story of an arroyo full of bones. It took them another couple of years to mention the find to someone who cared: Figgins, who enthusiastically embraced their discovery in 1926. Figgins and Harold Cook visited the site in March and decided that it was worth returning in the summer to excavate. They started digging in May 1926 with the intent of collecting a skeleton they could mount for display. In July, the workers found a spear point, but it wasn't in place. Now Figgins knew he was on to something, and he urged the workers to proceed with extreme caution. On August 27, 1927, they found an unusually fine fluted spear point lying in place in a spray of Ice Age bison ribs. Figgins ordered Schwachheim to guard the point and let no one touch it. Schwachheim duly sat down and awaited the arrival of "Scientists, Anthropologists, Zoologists, or other bugs."

Figgins wired New York, and, as luck would have it, Barnum Brown, the great dinosaur finder and the American Museum's number-one field man, was in Grand Junction. Brown joined Figgins at Folsom on September 4, 1927, and confirmed that Figgins was right. Humans had been in North America at the end of the Ice Age and they had been hunting the big animals. The advent of carbon dating showed the site to be 10,500 years old. A few years later, Figgins followed this discovery with a mammoth kill site at a place near Denver called Dent.

Mosasaur.

Kay showed us the Folsom point replicas that they had on display, and we stuffed some cash in the donation box and left town feeling we'd just had one of the more authentic experiences of our whole trip.

We crossed back into Colorado through a tight, winding valley called Tollgate Canyon. The walls of the canyon are sandstone and occasional dark gray mudstone layers. Ray now recognized the 100-million-year-old Dakota Sandstone when he saw it. He was able to predict that the land would flatten when we popped out of the canyon and up into the marine shale, and moments later, we were speeding along on the tabletop plains of southern Colorado near Branson. The sky was classic Colorado blue and the paired snowcapped Spanish Peaks in the distance pulled us west, but we turned east to make our rendezvous with Bruce Schumacher.

After 90 minutes of straight, dry road and blasting U2, we pulled up next to Bruce's green forest-service pickup at the turnoff for Vogel Canyon. Ray had hosted Bruce in Ketchikan the previous year and had seen him eyeing his son Patrick's drum set. On this day, Bruce was sporting a ponytail, scraggily goatee, blue jeans, and a black T-shirt without a single dinosaur reference. "Finally, a hip federal employee," said Ray. As we hopped out of the truck, Bruce loped over and welcomed Ray with the phrase, "You're standing on plesiosaurs, *Xiphactinus*, and mosasaurs." A federal employee who knew what he was standing on. We got into Bruce's truck and headed off across the plains toward Purgatoire Canyon.

Called "Picketwire" by the locals who have no truck with French, Purgatoire Canyon was explored by Spanish conquistadors in the 16th century. A number of the Spanish soldiers had the misfortune to die without the benefit of attending clergy, and their souls were thus condemned to purgatory. The canyon was thus given the name El Rio de las Animas Perdidas en Purgatorio ("the river of lost souls in purgatory"). French trappers changed it to Purgatoire. The canyon is there because the Purgatoire River cut down through the thousands of feet of soft marine shale and chalk and then sawed into the much tougher and more resistant Dakota Sandstone and the underlying Morrison Formation. On most of the rest of the plains of southeastern Colorado, the Dakota lies deeply buried beneath the soft marine sediments. So soft are they that they weather flat, forming the monotony of the landscape that keeps most Coloradoans oriented toward the mountains. From a fossil lover's perspective, the eastern plains are one huge sushi platter, literally covered with the remains of marine life that lived here during the late Cretaceous.

Bruce did a doctorate on the Cretaceous marine rocks of South Dakota, and Ray is certifiable when it comes to Cretaceous marine creatures, so the conversation went submarine fast. The layer directly above the Dakota is the Graneros Shale, and it has produced skeletons of

Brachauchenius, a horrifically big pliosaur with a head the size of a sperm whale. We weren't too far west from Pritchett, where Harvey Markman collected a nearly complete long-necked plesiosaur in 1939. A WPA worker named Fred Roth had found the 45-foot-long skeleton weathering out of a stream bank. Locals came from all around to see the beast, and many of them collected souvenirs. By the time the local high-school principal finally wrote the museum to report the find, pieces were scattered around two counties. The first step in Markman's excavation was to visit all the souvenir collectors and retrieve the scattered bones.

The specimen was described and named by Sam Welles, who named it after the president of the Denver Museum board, Charles Hanington. *Thalassomedon haningtoni* is an animal I see every day, because we have two casts of this graceful swimmer hanging from the ceiling of the entry atrium of the Denver Museum of Nature & Science. Visitors often peer up at the skeleton and remark that its shape is odd for a flying animal. Long-necked plesiosaurs are the real beasts that loaned their form to the imaginations of those who would claim to see the Loch Ness and other lacustrine monsters around the world.

Picketwire Canyon is a very well-kept secret. For many years, the army used it for training exercises, and they still use some of the surrounding area, so tank tracks and 50 mm cartridges from airborne warthogs litter the landscape. Because people have been excluded for so long, most people don't even know that southeastern Colorado has its own amazing canyonlands. The army transferred a big part of the canyon to the forest service in 1992, and the service began the long process of evaluating their new resource.

With cholla cactus, coyotes, and roadrunners, the canyon seems like the set of a Warner Brothers cartoon. But a closer inspection shows that the canyon was occupied by early Native Americans, conquistadors, and Spanish settlers, and the rock walls are lathered with petroglyphs. Federal archaeologists are a special breed, and the assessment of the archaeological resource has been a labored process. The riches have convinced the forest service to keep public access to a minimum. When Bruce took the job, he was essentially walking into a lost valley of paleontology.

He and his team of volunteers have been surveying the fossils and beginning to excavate the best ones. They recently recovered a roadside *Allosaurus* with a belly full of 86 gastroliths, or stomach stones, a real anomaly for a carnivorous dinosaur. With hundreds of square miles of not yet prospected Morrison Formation, the future is bright for more finds. We drove for about five miles, stopping here and there to look at archaeological sites before we arrived at the most spectacular dinosaur footprint site in the country.

The Picketwire track site is a flat-lying sheet of limy sandstone in the Morrison Formation that forms the bed of the Purgatoire River for a couple of hundred yards. The river is continually sweeping the bedrock clean, then covering it, then ripping it up. The site is dynamic. Famous dinosaur tracker Roland Bird apparently found the site back in the 1930s before he got distracted by the Glen Rose track site in Texas. He lost interest, and the closure of the valley by the army pulled the curtains of obscurity over the site. It wasn't until the forest service started to survey the valley in the early 1990s that the site came back to light. Martin Lockley studied it, and his fame was magnified when Louie Psihoyos photographed him at Picketwire for *National Geographic.*

The Purgatoire River was running fast but not deep, so we waded the channel in our bare feet. Because the channel bottom was smooth sandstone, there was no problem with big cobbles. "But," Bruce said, "watch out for the dinosaur tracks. If you step in one you could go in up to your waist." With that, he slowly waded across the 30-yard channel. Ray followed him, and so did I. Once on the other side, the size and magnificence of the site became apparent.

A sauropod dinosaur foot is shaped like the bottom of a telephone pole, basically a big round cylinder, like the foot of an elephant. The exposed sheet of sandstone on the far side of the river was as wide as a two-lane road and perhaps a hundred yards long. The surface was

(right) From top to bottom *Nyctosaurus, Protosphyraena,* and *Thalassomedon* frolicking in the Cretaceous sea of eastern Colorado.

marked by many very clear sauropod trackways. Most of the animals had been walking in the same direction, and I had the sudden and powerful feeling that I was standing on a Jurassic Serengeti. Ray and Bruce paced alongside the trails while I shot photographs. Two things were immediately clear: the absence of tail-drag marks meant that sauropods kept their tails in the air, and the narrow gauge of the tracks meant they walked with their feet directly beneath their bodies. These were two insights of the Dinosaur Renaissance and here they were, literally cast in stone. The third piece that struck me hard was that all of the trails were going the same direction. It was at least possible that we were seeing evidence of a herd of animals moving together.

（right)
The Purgatoire sauropod trackway, about as clear as a fossil site could possibly be.

Ray marveled at the huge surface. By taking our time and moving slowly and gingerly in our bare feet, we began to notice the smaller three-toed tracks of a medium-sized theropod. These sauropods had not been alone. *Allosaurus* or some other large three-toed predator had walked on the same lakeshore, perhaps with not the best of intentions.

(opposite)
Xiphactinus on a dry fly.

Wading back across the river, I stepped into a submerged dinosaur track and stumbled into thigh-deep water. I'd been warned, so I couldn't complain about my jeans being soaked. Bruce took us to one last site, a place where he and his volunteers had, on the last day of a survey, found the edge of a nice big *Apatosaurus* skeleton. "How many hundreds of sauropod skeletons are there in this valley?" mused Ray.

As we drove out of the canyon, Bruce put Jimmy Cliff in the CD player and we rolled back up to army training grounds with Jimmy singing, "Sittin' Here in Limbo." I'd made dinner plans with my wife and a couple friends who wanted to meet Ray, so we knew that this trip would end on schedule. It's at least 200 miles from Lamar to my condominium in Denver's Capitol Hill. We broke free from Bruce's Lost Valley of Paleontology around 3:00 P.M. Dinner was scheduled for 7:30, so we had to hoof it. Fortunately, the roads of eastern Colorado are straight, flat, and empty. I set my bearings for the quaintly named town of Punkin Center and let 'er rip. Ray played with the video feature of his digital camera and we conspired to prevent this endless road trip from ending. The sun was out and it felt fantastic. The sky was beginning to light up with a classic Denver Bronco sunset: blue sky and orange clouds. I remarked, for perhaps the hundredth time, that we were driving over the top of the Pierre Shale and that ammonites surely were to be found beneath us. Ray pulled out a five-dollar bill and said, "Prove it." I rolled the truck to a stop next to a small road cut and walked across the road with my hammer in my hand. The cut wasn't much taller than the truck, and it was almost completely grassed over. I picked my way through the roadside trash and climbed to the crest of the cut. A blazing red sun was lying right on the horizon and shining directly into my eyes, so I backed down the slope to get back into the shade. I spotted a piece of harder rock standing out against the weathered gray shale, picked it up, and split it with the hammer. One rock, one hammer blow, and one perfect pink, pearly ammonite. I uttered, "Got one" to Ray and walked over to where he was rummaging in the dirt.

"Dr. J," he swore as he handed over the fiver, "fossils really are everywhere."

KANSAS OCEAN LIFE

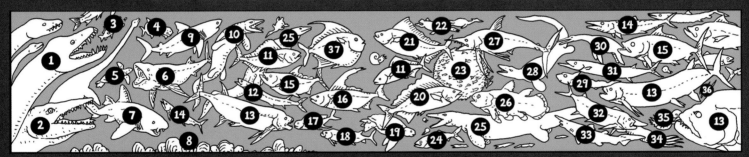

1. *Styxosaurus snowii*, elasmosaur
2. *Platecarpus tympaniticus*, mosasaur
3. *Kansius sternbergi*, bony fish
4. *Micropycnodon kansasensis*, bony fish
5. *Toxochelys latiremis*, turtle
6. *Bonnerichthys gladius*, bony fish
7. *Ptychodus mortoni*, shark
8. *Platyceramus platinus*, giant clam
9. *Cretoxyrhina mantelli*, shark
10. *Clidastes liodontus*, mosasaur
11. *Niobrara encarsia*, bony fish
12. *Saurocephalus lanciformis*, bony fish
13. *Xiphactinus audax*, bony fish
14. *Cimolichthys nepaholica*, bony fish
15. *Pachyrhizodus caninus*, bony fish
16. *Pentanogmius fritschi*, bony fish
17. *Thryptodus zitteli*, bony fish
18. *Gillicus arcuantus*, bony fish
19. *Protostega gigas*, turtle
20. *Pentanogmius evolutus*, bony fish
21. *Bananogmius sp.*, bony fish
22. *Hesperornis regalis*, diving seabird
23. *Apsopelix anglicus*, bony fish
24. *Enchodus gladiolus*, bony fish
25. *Tylosaurus proriger*, mosasaur
26. *Megalocoelacanthus dobiei*, bony lobe-finned fish
27. *Protosphyraena perniciosa*, bony fish
28. *Plesioplatecarpus planifrons*, mosasaur
29. *Edaphodon laqueatus*, ratfish
30. *Dolichorhynchops osborni*, short-necked plesiosaur
31. *Saurodon leanus*, Bony fish
32. *Ichthyodectes ctenodon*, bony fish
33. *Scapanorhynchus raphiodon*, goblin shark
34. *Niobrarateuthis bonneri*, giant squid
35. *Prionotropis hyatti*, ammonite
36. *Baculites sp.*, baculite ammonite
37. *Dixonogmius sp.*, bony fish

CODA

It was fitting that we ended this trip on the bed of the Cretaceous sea floor, more or less where we had started it some seven years before. Now another 18 years have passed, and both of us are getting closer to being fossils ourselves. After 30 years of bustling T-shirt retail on Creek Street, Ray and Michelle are planning to close the Soho Coho at the end of the 2023 tourist season. They recently bought a building in Lindsborg, Kansas, the town where Ray went to college and a place that calls itself "Little Sweden." The building will soon become the Prairie Sea Gallery.

It still blows my mind that the heartland of America was once the bottom of the sea.

Every Roadcut Asks a Question

ACKNOWLEDGMENTS

This road trip could not have happened without the support and cooperation of a host of enablers, friends, fossil fanatics, and family members. Ray thanks his wife, Michelle, and their kids, Patrick and Corinna. Kirk thanks his wife, Chase DeForest. Russ Graham authorized the use of Big Blue and some of Kirk's museum time for the trip. Jim Curtis kept Big Blue running. Both Ray and Kirk mourn the loss of people whom they encountered along the way who did not live to see the book published: Freida Gunther, Dave Love, Donna Engard, Bob Akerley, Wes Wehr, and Bill Bateman. Big Blue herself died in the spring of 2006. Other losses include Denver's Walnut Café, which closed before its pancakes could become famous, and the famous Douglass Quarry at Dinosaur National Monument, which closed indefinitely due to a faulty foundation.

This book was originally acquired for Fulcrum by Marlene Blessing and shepherded into print by Faith Marcovecchio, Sam Scinta, and Ann Douden. The text was improved by comments from Bruce Schumacher, Peter Larson, Dina Venezky, Doug Nichols, Peter Heller, Ken Carpenter, Greg Wilson, Kirsten and Dick Johnson, and Chase DeForest. We thank Michelle Williams for the editing and Jon Hahn for all the updates to the layout for the second version with Chicago Review Press.

There would have been no story if it weren't for the thick mesh of fossil finders, watchers, and keepers that covers the western landscape like a prehistoric national guard. In rough order of their occurrence in the text, we thank Chuck Bonner, Barbara Sheldon, Les Robinette, John Shinton, Russ Graham, Emmett Evanoff, Brent Breithaupt, Mike Lewis, Dave Schmude, Chris Weege, Kelli Trujillo, Bob Bakker, Kent Hups, Gary Staab, Lisa and Stan Icenogle, Dave Brown, Bill Wahl, Russell Hawley, Lee Campbell, Julia Sankey, Peter and Neal Larson, Bob Farrar, Mike Triebold, Bill Stein, Dean Pearson, Tyler Lyson, Patty Perry, Marshall Lambert, Nate Murphy, Darren Tanke, Phil Gingerich, Kirby Siber, Burkhart Pohl, Dave and Jane Love, Anna Moschiki, Mike Kinney, Tom Rush, Renée Askins, Susan and Mayo Lykes, Jay Muir, Wally Ulrich, Tom Lindgren, Rick Hebden, Vince Santucci, Charlie Love, Steve Sroka, Sue Ann Bilbey, Lace and Jim Honert, Randy Fullbright, Carol McCoy Brown, Ann Elder, India Wood, Mike Graham, Bill Bateman, Frank Rupp, Bob Akerley, Dick Dayvault, Rob Gaston, Jennifer Schellenbach, Lin Ottinger, Dwayne Taylor, Scott Sampson, Mark Loewen, Celina and Marina Suarez, Mike Leschin, the Gunthers (Lloyd, Val, Freida, DeEsta, Glade), Don Tidwell, Cliff Miles, Robert Harris, Buck-a-Bug Jimmy Corbett, Toni and Nancy Clare, Herb Meyer, Donna Engard and Pat Monaco, Alex Downs, Josh Smith, Tom Williamson, Matt Celesky, Alfred Newkirk, Kay Thompson, and Bruce Schumacher.

Kirk also thanks those geologists and paleontologists who have trained him and welcomed him to their western field camps, fossil sites, and quarries: Sid Ash, Ed Belt, Bill Bonini, Tom Bown, Robyn Burnham, Ken Carpenter, Bill Clemens, Bill Cobban, Phil Currie, Erling Dorf, Dave Fastovsky, Bob Giegengack, Phil Gingerich, Russ Graham, Leo Hickey, Jack Horner, Steve Manchester, Hans Nelson, Doug Nichols, Gomaa Omar, Pete Palmer, Reuben Ross, Richard Stucky, Don Tidwell, Wes Wehr, Peter Wilf, and Scott Wing. He also thanks Kelsey Martin for causing the rock problem in the first place.

Ray and Kirk toiled over the big map for nine months and are grateful for help from John Alroy, Howard and Darlene Emry, Larry Martin, John Hoganson, Matt Celesky, Chuck Bonner, Clark Markell, David Elliott, Russell Hawley, Bob McChord, Ken Carpenter, Jerry Smith, George Stanley, Neal Larson, Ron Eng, and Jim Baichtal. Fellow Alaskan artist Terry Pyles added the glorious digital color.

Ray would like to thank Marjorie Leggitt for the original leaf composition on *Leaves of the Apocalypse* on page 73.

INDEX

READY TO CRUISE
THE FOSSIL FREEWAY YOURSELF?

Let Ray and Kirk's road map be your guide. This masterpiece is filled with hundreds
of prehistoric images based on actual fossils found across the West and just waiting to be discovered.

Available from ipgbook.com.

Need another road trip route?

Cruisin' the Fossil Coastline

Available at ipgbook.com

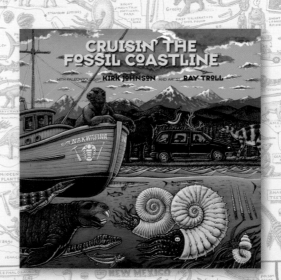